July, 1969. A *n:*
Aldrin and A *ıds*
collecting lun

GW01340141

Armstrong: 'Wh ' s
all I want to kn
Mission Control: 'What's there? . . . (garble) Mission Control calling Apollo II . . .'
Apollo II: 'These babies were huge, sir . . . enormous . . . Oh, God, you wouldn't believe it! I'm telling you there are other spacecraft out there . . . lined up on the far side of the crater edge . . . they're on the moon watching us . . .'

This transcript from Apollo II's moon mission is just one of many to which NASA refuses to give its official recognition. Here, at last, is the complete uncensored story behind the moon landings: *clear and indisputable facts offered by astronomers and the astronauts themselves.* This book also reveals NASA's private admissions about how and why America's powerful Space Agency remained silent in the face of such overwhelming evidence of UFO activity on and around the moon.

SECRETS OF OUR SPACESHIP MOON is the stunning sequel to OUR MYSTERIOUS SPACESHIP MOON: together these books represent an astonishing break-through in our understanding of our nearest space-neighbour.

Also by Don Wilson in Sphere Books:
OUR MYSTERIOUS SPACESHIP MOON

SECRETS OF OUR SPACESHIP MOON

Don Wilson

SPHERE BOOKS LIMITED
30/32 Gray's Inn Road, London WC1 8JL

First published in Great Britain by Sphere Books Ltd 1980
Copyright © 1979 by Don Wilson
Published by arrangement with *DELL PUBLISHING CO. INC.,*
NEW YORK, N.Y., U.S.A.

Grateful acknowledgement is made for permission to reprint the
following copyrighted material.

'Did Our Astronauts Find Evidence of UFO's on the Moon?' by
Joseph Goodavage: Reprinted by permission of *Saga* Magazine,
© Gambi Publications, Inc.

Excerpts from THE EXPANDING CASE FOR THE UFO'S by Morris
Jessup: Used by permission of Citadel Press.

From SOMEBODY ELSE IS ON THE MOON by George Leonard:
Copyright © 1975 by George Leonard. Reprinted by permission of the
David McKay Company, Inc.

From THE VOYAGES OF APOLLO by Richard Lewis: Copyright ©
1974 by Richard Lewis. Reprinted by permission of Quadrangle/Times
Books.

From OUR MOON by H.P. Wilkins: Published by Frederick Muller
Ltd., London. Used by permission of Anthony Sheil Associates Ltd.

Excerpts from *The New York Times*, Nov. 9, 1969 by Walter Sullivan
© 1969 by The New York Times Company. Reprinted by permission.

From 'The Moon Is More of a Mystery Than Ever' by Earl Ubell.
© 1972 by *The New York Times Magazine,* April 16, 1972.

Chart, 'The Speed of Sound Travelling Through Lunar Metals':
Reprinted with permission from CRC HANDBOOK OF CHEMISTRY
AND PHYSICS, 54th Edition. Copyright The Chemical Rubber Co.,
Inc., CRC Press, Inc.

TRADE MARK

This book is sold subject to the condition that
it shall not, by way of trade or otherwise, be lent,
re-sold, hired out or otherwise circulated without
the publisher's prior consent in any form of
binding or cover other than that in which it is
published and without a similar condition
including this condition being imposed on the
subsequent purchaser.

Printed in Canada

SECRETS OF OUR SPACESHIP MOON

The more you study the Moon, the more you will become aware that it is an orb of mystery — a great luminous cyclops that swings around the Earth as though it were keeping a celestial eye on human affairs.

— Frank Edwards

... the origin and history of the Moon have remained a mystery despite intensive study by eminent scientists during the last century and a half.

— Dr. Harold Urey

Nobel prize-winning scientists.

Of the bases the visitors might have from which to conduct their explorations of Earth, none holds more logical or intriguing possibilities than our Moon.
—*John Magor, editor,* Canadian UFO Magazine

INTRODUCTION

You are now about to embark on an intriguing intellectual odyssey. You are about to read one of the most fantastic theories ever conceived.

Some will regard this theory and its defense as bordering the lunatic fringe, or just simply science fiction. Others will immediately classify it as crackpot science. But people with open minds will, we believe, give it fair consideration. All we ask is that you let the evidence speak for itself.

For this book is constructed from a fabric of fact, not fiction. Scientific evidence supporting the Soviet theory that follows is taken from authenticated sources, much of it from the files of our space agency scientists. Proofs from the past as well as the present, cast interesting light on the origin and make-up of the Moon which has even modern-day experts baffled. Ancient historical documents reinforce a Soviet theory which explains the puzzling nature of this strange world in our skies. Proof positive will be forthcoming that the Moon is in fact not a natural satellite of Earth!

Do not reject the "unnatural" origin of this seemingly natural celestial body that circles us, just because it may seem at first sight to counter common sense. Five hundred years ago medieval scientists rejected Copernicus's crazy idea that our Sun and not the Earth was the center of our solar system universe. After all did not our eyes every day show us the truth—that the Sun rising in the east, traveling across the sky and setting in the west, absolutely prove that our star revolved around us. And not the world around the Sun! Copernicus was crazy, his contemporaries concluded.

However, common sense and the accepted scientific belief of his day were wrong. In time an "established" fact of science that was taught for hundreds of years was toppled into the dust. Nevertheless, the "crazy" priest Copernicus did not even dare to publish his heliocentric theory until he was on his deathbed, for fear of punishment which might include being burnt at the stake for his teachings!

Galileo was imprisoned and nearly put to the torch when he declared that the heliocentric theory was fact. Galileo too was thought to be a crackpot by many of his own contemporaries. He claimed that he had seen craters on the Moon and spots on the Sun through a new fangled gadget called a telescope. Impossible! Everyone at the time knew that the Sun was perfect!

We also know how other theories like those of Newton, Darwin and Freud were opposed by the scientific establishment of their day.

Even Harvey's discovery of the circulation of blood was laughed at. And Pasteur's idea that invisible germs or "little beasties" caused diseases was ridiculed and Pasteur considered a laughing stock by his scientific colleagues.

On and on the story runs through the entire history of science. In fact it would almost seem to be a rule of the history of science that what was considered unorthodox yesterday is now the accepted orthodoxy of today.

We have come over eons of time to look upon the Moon as a natural satellite of our planet. The unorthodox theory of two very orthodox scientific researchers of the Soviet Union that the Moon may in reality be a huge hollowed-out and internally reconverted "spaceship" may indeed someday become the orthodox truth of tomorrow.

In the twentieth century man has reached out into space and touched the Moon. He has walked and worked on an alien world—the first undoubtedly of many to come. In the entire catalog of human achievements man's landing on the Moon is, in our opinion, and in the opinion of many, among his greatest accomplishments. It certainly must be classified as man's greatest exploratory effort.

It may also lead to man's greatest discovery. For the Moon may, in fact, be the key to many earthbound mysteries, not the least of which is the mystery of man himself!

For another extraordinary story lies behind this peak of human achievement that demands telling. Another tale is buried here that remains largely unknown—not only of the strange seemingly artificial structures on the Moon that our space probes have discovered, nor the strange encounters of American astronauts with unidentified flying objects and an unknown alien presence around the Moon. Even more shocking is the largely untold story of the true nature of this mystifying mystery world that is orbiting our planet Earth!

If the theory of two Soviet scientists is correct and our interpretation of the brilliant work of American and international experts from all over the world who have practically torn the Moon apart scientifically is right, it could be the most shocking and important story of our times.

We therefore dare you to read this book and *open-mindedly* weigh the evidence. We honestly feel that you too will then come to agree with us that the Soviet theory is the most startling discovery of your life.

Many phenomena observed on the lunar surface appear to have been devised by intelligent beings. Now, U.S. and Russian moon probes have photographed two such "constructions" at close range.
—Dr. Ivan Sanderson

ONE

STRANGE STRUCTURES DISCOVERED ON THE MOON

Before man journeyed to our Moon many men of Earth were intensely interested in the possibility of finding evidence of alien beings having visited our satellite. Joseph F. Blumrich, a leading design engineer at the Marshall Space Flight Center in Huntsville, Alabama, concluded after an eighteen-month study on this subject that the Earth and the Moon have been visited regularly by extraterrestrial beings. He claimed that eventually we would find their artifacts on the surface of our lunar neighbor.

Why, despite half a dozen manned trips to the Moon, haven't we? To begin with, the area of the Moon's surface we have explored in six Apollo manned missions is minuscule. Another problem in our search for such artifacts is that they could be placed anywhere and might be in forms we have never imagined.

Science reporter Joseph Goodavage recently quoted a high-ranking General Dynamics Corporation executive who worked closely with NASA on the many Moon projects as stating: "An object or artifact placed by an alien civilization on the Moon could be something as obvious as a small pyramid sitting atop a mountain peak. We'd literally have to stumble over it before anybody would recognize it for what it is." He added: "But nothing less than a full-scale exploration of the Moon will turn up whatever is there." (*Saga*, April 1974.)

Probably such exploration will not take place until the twenty-first century at the earliest.

Actually, even without full-scale exploration, we have discovered a number of structures which appear to be artificial, during the few manned and unmanned space probes we have sent to our Moon. In fact, of the many mysteries uncovered in our journeys to this neighboring world, none is more mystifying than that of the strange structures discovered there.

In our book *Our Mysterious Spaceship Moon* (Dell, 1975) we discussed many different discoveries that have been made on the Moon—both American and Soviet. The amazing thing is that, as startling and seemingly important as they are, little is known of them by the public.

In fact, even before man journeyed to our satellite, reports of structures sighted on the surface of the Moon were numerous. Astronomical literature is crammed with them. One of the most startling is the huge bridgelike structure seen over the Sea of Crisis in 1954 by John O'Neill, former science editor of the New York *Herald Tribune*. This oddity was confirmed by other leading astronomers who also saw it through their own telescopes. Some estimated it to be 12 miles long.

Of course, the key question is, was it a natural structure or of artificial construction? The eminent British astronomer H. P. Wilkins, head of the Lunar Section, British Astronomical Association, made this startling statement in a BBC radio program: "It looks artificial." (Donald Keyhoe, *The Flying Saucer Conspiracy*, Holt, 1975.)

When he was asked what he meant exactly by the term "artificial," Dr. Wilkins answered: "Well, it looks almost like an *engineering* job." This lunar expert added that it was more or less regular in outline, and even cast a shadow under a low Sun. He startled everyone when he exclaimed: "You can see the sunlight streaming in beneath it."

In the entire radio interview, not once did Wilkins refer to this structure as being a "natural" bridge. Instead, he used words which indicated that he thought it might be artificial. The fact that the bridge had not been seen before, although the area was well known and often studied, increased the possibility that it might indeed be a construction made by beings not of this world—and fairly recently.

Many other seemingly intelligently constructed "structures," like walls that formed squares or rectangles, and even strange domelike structures that appeared out of nowhere and at times seemed to disappear, led many observers to believe that they are the work of alien intelligence.

The chances are, admittedly, that many apparently artificial constructions are nothing but natural formations, misinterpreted by Earth observers due to the vast distance between us and our nearest neighbor. Space photos of Earth teach us a lesson here. From outer space, orbiting vehicles have taken photographs that lend to many natural Earth formations an aura of artificiality. For instance, from the depths of space the Barringer crater of Arizona looks like an artificial construction. Similarly, photos of an area at the southern edge of the Sahara Desert in northern Nigeria show what appears to be a sprawling series of structured walls—but we know that these are merely natural formations. So man must be extremely careful in his conclusions.

PRE-APOLLO LUNAR MYSTERIES

Before our space program sent probing mechanical eyes to the Moon, throughout the many decades and even centuries of the so-called telescopic age of lunar study, competent observers on Earth saw strange things on our satellite: unexplainable weird lights and glows sighted on the surface, inexplicable changes, and the sudden appearance of structures. Astronomers thought that the Moon was a lifeless, airless, windless, and for the most part erosionless world. In fact, a completely changeless orb. Yet unquestionably many changes did take place.

For instance, in 1843 Johann Schroeter recorded unexplainable changes in a six-mile crater named Linne. This German astronomer made hundreds of maps of the Moon over many years. Over a lifetime of observations, the crater Linne gradually disappeared. Today Linne is just a tiny bright spot with little depth or height, a small pit surrounded by whitish deposits. What happened? No one knows for sure.*

---

*It should be noted that NASA photographs taken by Apollo 15 reveal that Linne today is a tiny crater (1½ miles across). The mystery of what astronomers observed has remained just that—a mystery.

Could this competent astronomer have been mistaken? Schroeter himself believed there was intelligent life on the Moon, and he attributed some of these changes to "the industrial activities of the Selenites."

Although Schroeter's contemporaries heartily attacked his hasty conclusion that the Moon was inhabited by intelligent beings, nevertheless, in Schroeter's day, unlike our times, many scientists did consider the Moon to be an inhabited world as is Earth.

The great British astronomer H. P. Wilkins, commenting on these unexplainable changes and Schroeter's beliefs, makes this striking observation:

> We cannot subscribe to this idea because without air to breathe it is exceedingly difficult to contemplate the existence of Selenites let alone to speculate as to their possible activities, industrial or otherwise. It is equally difficult to explain these things on natural grounds.
>
> But there remain some still greater puzzles. Some objects do not vary in their physical features so much as to the tints or hues of their interiors. There are craters which change colour in a very peculiar manner. What looks like a green carpet can be seen spreading over their floors! (H. P. Wilkins, *Our Moon*, Frederick Muller, Ltd., 1954, p. 130.)

Even more wild than Schroeter's speculations were those of his fellow countryman and astronomer Gruithuisen, who was convinced that great cracks on the Moon were actually canals or roads. Wilkins wryly notes: "It was just before the first railways were built, otherwise he would probably have said they were railways." (*Our Moon*, p. 57.)

Even more bizarre were reports of other competent astronomers. On the night of July 6, 1954, Frank Halstead, former curator at the Darling Observatory in Minnesota, along with his assistant and sixteen visitors, observed a straight black line in the crater Piccolomini, where none had been detected before. It was to disappear shortly thereafter, although not before other competent astronomers confirmed the discovery.

SIGNS ON THE MOON?

One of the strangest of all such lunar reports comes out of Japan, where *Mainichi*, one of Japan's largest newspapers, reported the unusual discovery of Dr. Kenzahuro Toyoda of Menjii University, who, while studying the Moon through a telescope on the night of September 29, 1958, spotted what appeared to be huge black letters, so pronounced they were easily discernible. The letters seemed to form two words: PYAX and JWA. No one to this day knows what these letters seen on the Moon mean or can give an explanation to the experience.

We emphasize that these are the kind of reports made by observers whom Wilkins, former director of the Lunar Section of the British Astronomical Association, calls "people who have been observing the moon too long to be easily deceived."

Dr. Wilkins adds: "In any case it is incredible that all were the victims of hallucination. We must accept the records even if we cannot explain them. Our knowledge and opinions are the products of existence on the earth; it is reasonable to suppose that on another planet conditions exist and events take place which have no counterpart on our planet. If there are intelligent creatures on other worlds, they are most unlikely to have bodies, or minds like ours; it would be very surprising if they did."

Wilkins concludes: "The moon is an alien and foreign world and much of what happens up there must remain a mystery until men actually land on its warty surface." (*Our Moon*, p. 139.) *

The objection might be made that man has gone to the Moon, photographed it completely at close range, and even landed and explored some of its surface. And no evidence of intelligently made structures or activities was uncovered. Or was it?

The surprising answer to this objection is that it was!

* The author is pleased to report that the British Astronomical Association has announced the formation of a special Lunar Section (of the NSW Branch) which will concentrate on solving the mystery of changes and unexplainable Lunar discoveries.

Photos, both American and Soviet, reveal that seeming non-natural, artificially made structures do exist on the Moon!

SOVIET DISCOVERY OF "MONUMENTS" ON THE MOON

The Soviet space probe Luna-9 took some startling photographs (February 4, 1966) after the vehicle had landed on the Ocean of Storms, one of those dark, circular "seas" of lava on the Earth side of the Moon. The photos revealed strange towering structures that appear to be lined up rather than scattered randomly across the lunar surface.

Dr. Ivan Sanderson, the late director of the Society for the Investigation of the Unexplained and science editor for *Argosy* magazine, observed that the Soviet photographs "reveal two straight lines of equidistant stones that look like markers along an airport runway. These circular stones are all identical, and are positioned at an angle that produces a strong reflection from the Sun, which would render them visible to descending aircraft." (*Argosy*, August 1970.)

But Sanderson was not the only reporter revealing these strange structures to the world. The Soviet press also carried articles on them. The Soviet magazine *Technology of Youth* gave an extensive report on them, calling them "stone markers" which were unquestionably "planned structures," and suggested that these "pointed pyramids" were not natural formations but definitely artificial structures of alien origin.

After examining the photographs of these objects, Dr. S. Ivanov, winner of the Laureate State Prize (which the Soviets consider equivalent to the Nobel Prize), calculated from the shadows cast by the spirelike structures that at least one was about fifteen stories high.

Ivanov, who is also the inventor of stereo movies in the Soviet Union, pointed out that by luck—perhaps the space probe landed on a spot where the ground had settled, or set down upon a small stone or rough spot—"a chance displacement of Luna-9 on its horizontal axis had caused the stones to be taken at slightly different angles." This double set of photographs allowed him to produce a three-dimensional stereoscopic view of the lunar "runway."

The result of this bit of good fortune, as Ivanov reports, was that the stereoscopic effect enabled scientists to figure the distances between the spires. They found, much to their surprise, that they were spaced at regular intervals. Moreover, calculations confirmed that the spires themselves were identical in measurement. Says Ivanov: "There does not seem to be any height or elevation nearby from which the stones could have been rolled and scattered into this geometric form. The objects as seen in three-D seem to be arranged according to definite geometric laws."

This discovery must be heralded as among the most important discoveries made by either the American or Soviet space program. But, strangely enough, for the most part they have been ignored. As we shall soon see, other discoveries, equally as important, have been covered up by our own space agency. In fact, Art Rosenblum, head of the Aquarian Research Foundation, who says he learned of the Soviet discovery from Lynn Schroeder and Sheila Ostrander, the authors of *Psychic Discoveries Behind the Iron Curtain,* before their work was published in America, claims they indicated that authorities at NASA "were not at all happy about its publication." Why not? What is NASA trying to hide? asks Rosenblum. (Arthur Rosenblum, *Unpopular Science,* Running Press, 1974.)

Another question that probably arises in most readers' minds is, what was the purpose of these structures, assuming that they were built by alien beings? Dr. Sanderson speculated: "Is the origin of the obelisks on the Earth and those on the Moon the same? Could both be ancient markers originally erected by alien space travelers for guidance of late arrivals?" He pointed out that it seems hard to understand why man ever started making obelisks anyway, since it is a very difficult job and seemingly purposeless. Or did obelisks have a purpose other than Earthly? Could these spirelike structures actually be signal spots for the coming and going of spaceships, as some speculated? Not marking the landing on outer Moon bases but for underground, hidden bases located inside the Moon?

Intriguingly, on the edge of this same Sea of Storms is a strange opening that leads down into the Moon. Dr. H. P. Wilkins, one of the world's leading lunar experts before his untimely death a few years ago, was convinced that extensive hollow areas did exist inside the Moon, perhaps in the form of caverns, and that these were connected to the surface by huge holes or pits. He discovered such an opening himself—a huge round hole inside the crater Cassini A. This crater is one and a half miles across, and the opening leading down into the Moon is over 600 feet across—more than two football fields laid end to end. Wilkins writes in his definitive work, *Our Moon:* "Its inside is as smooth as glass with a deep pit or plughole, about 200 yards across at the centre."

As we shall see in the coming chapters, hundreds, in fact thousands, of UFOs have been seen on or around the surface of the Moon, and a concentration of them has been spotted in the Sea of Storms. Could they be coming and going through this huge opening or one like it?

AMAZING AMERICAN DISCOVERY OF PYRAMIDAL STRUCTURES ON THE MOON

The Soviet discovery is mind-boggling enough. But, amazingly, a similar discovery was made by the American space probe Orbiter 2, which took pictures on November 20, 1966, from a height of 29 miles above the Sea of Tranquility—the very same area where our first astronauts landed on the Moon.

The photos of Orbiter 2 show what appear to be the shadows of several pointed spires shaped like obelisks—long needlelike structures similar to the Washington Monument and to Egyptian obelisks like Cleopatra's Needle, now in Central Park, New York.

From the shadows cast by these long, needlelike spires scientists have variously estimated that they range from 40 to 75 feet high. Soviet scientists examining the American Orbiter 2 photos claim that they are much higher—at least three times as high as the highest American estimate, which would make them as tall as a fifteen-story building! A few scientists, like Dr. Farouk El Baz (formerly one of NASA's leading geologists, now with the Smithsonian Institution), estimate that these spirelike structures on the Moon are as tall as the tallest buildings on Earth—and probably even taller! (*Saga,* March 1974.)

More important than their height or size, however, is their positioning. Dr. William Blair of the Boeing Institute of Biotechnology claims they are *geometrically* positioned. Blair, a specialist in physical anthropology and archeology, observes: "If the cuspids [these spirelike stone structures] really were the result of some geophysical event it would be natural to expect to see them distributed at random. As a result the triangulation would be scalene or irregular, whereas those concerning the lunar object lead to a basilary system, with co-ordinates x, y, z to the right angle, six isosceles triangles and two axes consisting of three points each." The *Los Angeles Times* (February 26, 1966) carried a drawing of Blair's geometrical analysis of the positioning of these spires, as they were photographed by Orbiter 2. (See the NASA photo on inside back cover, and note the long, pointed shadows which indicate seven spires. Hold the picture upside down for proper effect.)

Because of this peculiar geometrical positioning, Blair is convinced that these seven spires grouped closely together are not randomly placed. To him the spires are significant simply because they form a right-angled coordinate system, resulting in six isosceles triangles and two axes of three points each. He believes this cannot be happenstance. Furthermore, as Blair points out, there is also evidence that a large rectangular pit exists just west of the largest spire. As he observes: "The shadow cast by this depression seems to indicate four 90-degree angles and resembles the profile of an eroded structure."

Blair insists that they should be investigated more thoroughly, for, as he points out, if a similar thing had been found on the planet Earth, "archeology's first concern would have been to inspect the place and carry out trial excavations to assess the extent of the discovery."

As this expert anthropologist and scientist notes, if similar structures on Earth had been passed off in such a fashion more than half the Mayan and Aztec architecture known today would be "still buried under hills, depressions covered in trees and woods." In fact, concludes Blair: "If these had been passed off as a result of some geophysical event the science of archeology would have never been developed, and most of the present knowledge of man's physical evolution would still be a mystery."

However, not all scientists agree with Blair's assessment of these strange structures. Dr. Richard W. Shorthill of the Boeing Scientific Research Laboratory claims that "there are many of these rocks on the Moon's surface. Pick some at random and you eventually will find a group that seems to conform to some kind of pattern."

This is a gross exaggeration on Shorthill's part, for though other such strange groupings of needlelike spires and similar puzzling constructions exist on the Moon, they do not proliferate in such large numbers that chance, random arrangements of a distinct geometrical nature can be found all over the place. Furthermore, other scientists, such as Soviet space engineer Alexander Abramov, have examined the Orbiter photos and concur with Blair's judgment that they are indeed geometrically positioned. But Abramov believes they are positioned in a very unusual way. His geometrical analysis, made by calculating the angles at which "they appear to be set," leads Abramov to conclude shockingly that they form what is known as an "Egyptian triangle." Seemingly artificial constructions on the Moon that just happen to form what is known among archeological and historical experts on Earth as an Egyptian triangle?

Says Abramov: "The distribution of these lunar objects is similar to the plan of *the Egyptian pyramids constructed by Pharoahs Cheops* [the Great Pyramid], *Chephren, and Menkaura at Gizeh, near Cairo.* The centers of the spires in this lunar 'abaka' are arranged in precisely the same way as the apices of the three great pyramids." (*Argosy*, August 1970. Emphasis added.) Also see American photo of this strange positioning.*

If Dr. Abramov's calculations (as reported by Dr. Ivan Sanderson) are correct, then this is not only startling evidence of intelligence on the Moon but leads to the reasonable conclusion that this intelligence left its telltale marks on the planet Earth.

Only up to now we have not recognized them for what they are!

There is yet another clue which indicates these objects are not natural formations but artificial constructions. Dr. Farouk El Baz, who maintains that some of the spires are

---

* See NASA photo of these structures: Inside back cover.

"taller than the tallest buildings on earth as calculated by the tremendously long shadows they cast on the Moon's surface" (perhaps even more than two to three times the height of the tallest structures on Earth!), points out that these structures are of a much lighter color than the surrounding lava fields and landscapes, which indicates that they are "constructed of different materials." (*Saga,* April 1974.)

Such conclusions on the part of respected space scientists —both Soviet and American—indicate that what we have here is the first concrete evidence of the existence of an alien intelligence on the Moon.

But if we are to believe two Soviet scientists of the renowned Soviet Academy of Sciences, there is a whole world of evidence that the Moon *itself* indicates that not only was there intelligent life on the Moon at some indeterminate time in the past, but intelligent life has been living *inside* the Moon for eons. The evidence they have compiled leads them to believe that the Moon may be a hollowed-out spacecraft of a sort, steered into orbit around our Earth eons ago!

IS NASA COVERING UP THE ORBITER 2 DISCOVERY?

Art Rosenblum of the Aquarian Research Foundation claims he met a scientist formerly employed by NASA who had helped design the Houston Space Center.

Rosenblum declares: "He told me that at about the time of . . . photographing the shadows of the obelisks . . . the Boeing Aircraft Corporation published this same photo in their company newsletter. . . . He had not seen it published since then. He also told me that while working for NASA he found it uncommonly difficult to get information from them. . . ."

Rosenblum pointedly asks: "If monuments had been discovered on the Moon, one would suppose that that is about the most important result of the whole Moon probe effort. Why is this information not widely publicized by NASA? Why are they not fully investigated—or are they? What other type of information is being withheld? Why? Should not NASA be investigated?" (*Unpopular Science.*)

Rosenblum's implications are not quite true. Certainly NASA can be accused of cover-ups, as we have seen, but to assert that they never released any information or photos on this at all is totally inaccurate.

On November 22, 1966, NASA did release a photo (see photo, inside back cover). At the same time, NASA denied that it revealed anything—so very few news publications or other media picked up on the story. After all, hadn't *the authority*, the National Aeronautics and Space Administration, spoken?

However, the Washington *Post* and the *Los Angeles Times* were among the few that carried stories clearly showing that there might be something to the strange photo.

The Washington *Post*, in fact, ran a front-page story showing the startling photo with this headline:

6 MYSTERIOUS STATUESQUE SHADOWS PHOTOGRAPHED ON THE MOON BY ORBITER

Post staff writer Thomas O'Toole noted that the photo indeed showed "six shadows" which were "hailed by scientists as one of the most unusual features of the Moon ever photographed."

Measurements of the huge shadows in the photo, O'Toole went on to point out, showed them to be as short as 20 feet and as long as 75 feet. One scientist referred to these needlelike shadows on the Moon, says O'Toole in an offhand way, as "Christmas tree effect." Another referred to their "Fairy Castle" effect. However, one scientist was impressed and called the region the Moon's "Valley of the Monuments."

While NASA feigned ignorance as to just what could be causing these shadows, O'Toole noted that "the largest shadow is just the sort that would be cast by something resembling the Washington Monument, while the smallest is the kind of shadow that might be cast by a Christmas tree. (Washington *Post*, November 22, 1966.)

The New York Times took a much more conservative view of the startling nature of these strange shadows and the objects creating them. On November 24, 1966, Dr. Thor

Karlstrom of the U.S. Geological Survey was quoted by the *Times* as saying:

"The objects casting the shadows are not so nearly spectacular as the shadows themselves."

Karlstrom insists that a very low Sun (11 degrees) makes them appear much longer than they really are. He even denies that all have "a spire-like" appearance, although he does admit that these "shapes are very, very interesting." He says at least a couple of the shadows indicate that they are created by "squat blocks rather than spires because they appear wider than they are high."

Other scientists dispute Karlstrom's calculations and conclusions, as we have seen, but Soviet scientists who have investigated these photos generally agree that they reveal spires, all right, and much higher spires than even the most generous American scientists are willing to admit. In fact, the Soviets claim that not only are there towering spires but their unusual position gives away the fact that they are artificial in construction.

THE POSSIBLE TRANQUILITY CONNECTION

The strange, seemingly artificial structures discovered by Orbiter 2 show them to be located on the Sea of Tranquility. Was it by accident that the very first men landing on the Moon, on the very first manned-landing mission to the Moon, were sent to this very same area, near these mysterious pyramidal, obelisklike structures?

Was this just a coincidence? We do know that our space authorities at NASA knew of the existence of these striking structures. They were photographed in 1966, long before Apollo 11. Surely NASA, knowing of the existence of such strange structures and, contrary to their public pronouncements, not knowing for sure whether they were artificial or natural, would have been anxious to investigate these strange spires. Surely they did. But why is it that they did not announce their findings then? Furthermore, dare we ask: *Did NASA actually choose the Tranquility site because of the Orbiter 2 discovery?* Perhaps the answers to these important questions will become known. Perhaps someday we shall know the full truth.

We do know that Dr. Farouk El Baz, one of NASA's former scientists, now research director at the National Air and Space Museum of the Smithsonian Institution, admitted in a magazine interview that NASA actually did carry out secret investigations, and he insisted that "not every discovery has been announced." In fact, El Baz claims that NASA was "looking for something" on its manned Moon missions. Was that perhaps a much bigger artificial construction? For if the two Soviet scientists are correct, then the Moon itself is in part an artificial construction—a natural asteroid converted into a hugh hollowed-out spaceship!

Be prepared to face up to new explanations for old mysteries on the Moon!
—*George Leonard*

TWO
MYSTERIOUS LIGHTS ON THE MOON!

Not only have strange structures been discovered on the Moon, but even more puzzling, mystifying, unexplainable lights and moving objects have been seen on a world that scientists insist is completely dead. In fact, there have been so many reports of mysterious lights and unexplainable "happenings" taking place on our satellite that our own government space agency, the National Aeronautics and Space Administration, itself produced a study of them. Interestingly, this was done *before* this same government agency sent men to the Moon.

This remarkable study was called *Chronological Catalogue of Reported Lunar Events*. It included a listing of strange lights (both stationary and moving), glows, and "happenings" reported by reliable observers over the past several centuries. Doubtful reports have been excluded.

Although the study is extensive, it certainly is not complete. Another group of scientists, working independently, reported over 800 solid sightings of strange lights and glows on our Moon. Others claim that they actually number in the thousands. But the NASA study is impressive. It is impressive just to think that our own space agency took the trouble to make such a study. Surely this indicates they suspected something unusual has been taking place on our neighboring world.

What kinds of reports does the NASA study include? It is impossible, of course, to give even superficial coverage of the entire study, but we have included a fairly representative sampling.

Here is a selection of some of the more impressive sightings:

- On March 5, 1587, "a star is seen in the body of the moon . . . whereat many men marvelled, and not without cause, for it stood directly between the points of her hornes." (Harrison, 1876; Lowes, 1927.)

- November 12, 1671—A small whitish cloud sighted on the Moon by the well-known scientist Cassini. Since clouds do not exist on the Moon, what could it have been? Interestingly, today, as in ancient times, UFOs have commonly been described as "clouds." And in this NASA account "clouds" were reported to have been sighted on the Moon dozens of times.

- May 18, 1787—"Lightning" was seen on the face of the Moon by astronomers Halley and De Louville. De Louville explained them away as "storms." Neither "storms" nor "lightning" can take place in the airless world of our Moon.

- March–April 1787—William Herschel sighted three "bright spots" and an additional four "volcanoes" in April 1787. It is difficult to figure out what Herschel actually saw, since our lunar scientists have learned from Apollo studies that the Moon is a dead world and assure us that it has been volcanically dead for the past 3 billion years, and certainly there have been no volcanoes in recent times. What then were these "volcanoes" that the father of modern astronomy saw on the Moon?

Interestingly, Herschel reported that some of these strange lights seemed to be moving "above the moon."

• July 1821—The German astronomer Gruithuisen reported seeing "brilliant flashing light spots" on the Moon. "Blinking' 'or "flashing" lights are reported scores of times in this report.

• April 12, 1826—Black moving cloud over the Sea of Crisis (reported by Emmett). Interestingly, this is the same area where modern-day astronomers reported seeing a bridgelike structure suddenly appear in 1954, where none had been detected before. Is it coincidental that lights and other inexplicable lunar "events" have been reported in this same area dozens of times?

• February 1877—A fine line of light like "luminous cable" drawn west to east across Eudoxus Crater. The light was observed to last one hour. The average time "lights" lasted, according to the study, was over 20 minutes! They could hardly have been meteors flashing against the hard lunar surface, as some scientists claim.

• July 4, 1881—"Two pyramidal luminous protuberances appeared on the moon's limb . . . They slowly faded away . . ." What could this have been?

• April 24, 1882—Shadows, both moving and stationary, sighted in the Aristotle area. Moving shadows on the Moon? What could produce moving shadows except something moving? But what could possibly be moving on this dead world?

• January 31, 1915—Seven white spots arranged like a Greek gamma. What could this have been? Scientists do not know.

• April 23, 1915—A narrow, straight beam of light in the crater Clavius.

• June 14, 1940—Two hazy streaks of medium intensity, much complex detail. Seen in the crater Plato, where thousands of lights have been reported.

• October 19, 1945—Three brilliant points of light

on the wall of Darwin. Cited by Moore, one of the scientists who prepared the report.

• May 24, 1955—"Glitter," suggesting electrical discharge, sighted near the Moon's south pole. The well-known scientist Firsoff was the observer.

• September 8, 1955—Two flashes from the edge of Taurus. Coincidentally (?), this is where the astronauts of Apollo 17 were sent (Taurus-Littrow area).

• September 13, 1959—The area of Littrow was "obliterated by a hovering cloud." Could this have been a UFO?

• June 21, 1964—A moving dark area sighted by several observers in the area south of Ross D. It was observed for 2 hours 1 minute!

• July 3, 1965—Pulsating spot on the dark side of Aristarchus; seen for 1 hour 10 minutes.

• September 25, 1966—Blinking lights in crater Plato seen by several observers for minutes. Some described the lights as "reddish patches." Also seen the same day, red lights in Gassendi for 30 minutes. A month later (October 25) in the same place, "red blinks" were again seen by several astronomers.

• September 11, 1967—A "black cloud surrounded by violet color" was sighted in the Sea of Tranquility area (where the first mission to the Moon was to be sent) by a "Montreal group" of astronomical observers, according to this NASA report.

This is just a sampling of what is contained in this study. It was compiled by Jaylee M. Burley of the Goddard Space Flight Center, Patrick Moore of the Armagh Planetarium in Ireland, Barbara M. Middlehurst of the University of Arizona Planetarium, and Barbara L. Welther of the Smithsonian Astrophysical Observatory. Apparently NASA was impressed by the study—impressed that in fact these reliable reporters were actually seeing what they were reporting—for soon thereafter NASA carried-out Operation Moon Blink, a search for unexplainable lights and "happenings" taking place on the Moon. It was done in conjunction with cooperating observatories around the world, and in a short time Operation Moon Blink reported ten more such inexplicable lunar phenomena, three of which were confirmed independently and separately by observers outside the program. In fact, by August 1966 ten Moon Blink stations had detected twenty-eight lunar events! (*Lunar Luminescence,* Grumman Research Report.)

Of course, the key question is, what influence did the report and the subsequent Operation Moon Blink program have on the decision to send astronauts to the Moon? This was done *before* the Apollo missions took place, though certainly not before the decision had been made. We shall probably never know if it had any bearing, but the possibility is nonetheless intriguing.

The citings in this NASA report do not tell the full story. Frequently investigation reveals that the sighting was much more sensationalistic than the mere details included in the study indicate. Of course, this is understandable, for the final report includes only the barest facts.

For instance, consider an entry in this NASA-sponsored catalogue of lunar "events" which in fact describes a very unusual rash of sightings. The NASA study reads simply:

"No. 114—May 13, 1870. Location: Plato: Bright spots, extraordinary display. Observer Pratt, Elger; reported by the British Assn. 1871."

"Extraordinary" is an understatement! For the myriad of lights seen appeared in groups of as few as four and as many as twenty-eight! The lights were extraordinarily bright —in fact, extremely intense. They also seemed to pulsate at times: One would increase in intensity while others diminished. Almost as if—as one observer put it—"responding to the touch of switches of some mysterious lunar operator of electric batteries of lights!"

Thus, this tame entry actually covers up one of the weirdest and most inexplicable series of sightings ever recorded in the annals of astronomical lunar history. But the NASA report passed over it without mentioning its extraordinary nature. For lights were seen not only in the crater Plato but in the Sea of Crisis area. And not only just in

1871 but for several years reports of these strange lights and "events" in these areas came pouring in. The Moon continued to break out in a splurge of mysterious lights; in fact, they seemed to appear in such regular patterns that the Royal Astronomical Society of Great Britain held a special three-year investigation.

Again many of these strange, puzzling lights were spotted in the Sea of Crisis region—the very same area in which many astronomers had reported seeing a huge "bridge." The lights continued to appear again and again—sometimes singly, sometimes in groups; at times in straight-line formations, sometimes in circular or even triangular formations. Most seem to be moving or varying in intensity. Indeed, it appeared to some observers that they were under intelligent control!

In fact, though the Royal Astronomical Society would not admit it publicly, it is reported that privately many of its members expressed the belief that an unknown race af alien beings on the Moon were attempting to signal Earth.

It is estimated that about 2000 strange, mysterious lights were observed in this extraordinary three-year display.

Then, as suddenly as they appeared, they disappeared. What they actually were, no one has been able to figure out.

Something definitely strange is taking place on our neighboring world. Moving lights, unexplainable objects, seemingly artificially constructed and placed structures— all sighted on a supposedly dead and uninhabited world.

But there is more! The NASA study is far from complete. In our book *Our Mysterious Spaceship Moon* we detailed a whole host of impressive unexplainable lights and "events" on the Moon, seen by reliable reporters. Furthermore, Dr. William Corliss, formerly with NASA, has also compiled an impressive account of weird lunar lights and "happenings," most of which were not included in the somewhat conservative NASA report. His are also drawn from respected scientific sources.

Corliss writes: "Transient lunar phenomenon . . . has been observed ever since the invention of the telescope (and sometimes even without this instrument). It is impossible to reproduce the thousands of reports, and a representative sampling must suffice." (William Corliss, *Strange*

Universe, Custom Copy Center, 1975, AOL-103, A1-115.)
Here is a cross-section sampling of Corliss's collection, taken from leading scientific journals and writings of the past two centuries.

• William Wilkins noticed what he called a "star" passing over the Moon—which in the next moment he "realized was impossible. . . ."
On another occasion Wilkins even saw lights detaching themselves from the Moon. He observed these fixed steady lights for more than 5 minutes.

• Robert Hart, competent observer of the Royal Astronomical Society, observed "two luminous spots" of a "yellow flame colour," so bright that "they showed *rays around them as a star would do.*" (Emphasis added.)

• Johann Schroeter, a well-known astronomer, saw a point of light in the lunar Alps regions, as bright as a star, which disappeared only after he watched it for 15 minutes! To Schroeter's surprise, "where the light had been, a round shadow [now appeared] on the surface of the Moon, which was sometimes gray, sometimes black."

• Professor Holden, director of the Lick Observatory, on July 15, 1888, reported that he saw an "extraordinarily and incredibly bright [light] . . . the brightest object I have ever seen in the sky . . . ten times as bright as the neighboring portions of the Moon's surface."

This intense brightness was difficult to account for, although Holden thought it was only a volcanic eruption. Some of the more conservative explanations claimed that bright spots like this were simply "the sun reflecting off lunar snow." The only problem with this inaccurate explanation is the fact that snow does not exist on the Moon! Another attempt at a "natural" explanation maintains that they are merely reflections off metallic elements in lunar mountain peaks. But the phenomenon that Holden observed "blazed with such a dazzling brilliancy that it would be difficult to account for" it by such a flimsy explanation.

Others have gone to the opposite extreme and come up with bizarre explanations, such as the Sun shining off "metallic dikes" or "tremendous crystalized masses, with polished surfaces, throwing back the glare of the sunshine like mirrors."

However, critics point out that if this were the case the "glittering eminences" on the Moon would be nothing less than "enormous quartz crystals, whose dimensions are measured by miles instead of inches."

> • *Science* magazine (August 9, 1946, p. 146) also carried a report of lightninglike phenomena observed on the Moon; observers saw "some flashes of light streaking across the dark surface."
>
> • *Nature* magazine (August 18, 1887, p. 367) carried a report of "small cumulous cloud observed a little distance from the moon."
>
> • During a lunar eclipse strange fingers of light were seen "illuminating the upper section which was in shadow."
>
> • Dr. Frank B. Harris reported seeing the sudden presence on the Moon of "an intensely black body about 250 miles long and fifty wide. . . ." Harris said the sight resembled a crow poised on the Moon. "I cannot but think that a very interesting and curious phenomena (sic) happened." That is the understatement of the century!

It is interesting to note that these reports are missing from the NASA study. However, it is understandable why the compilers of the NASA technical brief left this out. How do you explain such a huge object over the Moon?

But this is just the problem. How do we solve these sightings? Before we traveled to the Moon the lights reported to have been seen there were commonly passed off as "volcanic eruptions." Now, however, since NASA scientists have learned that the Moon has been volcanically dead for eons, this explanation seems to be out. A NASA publication, *Apollo 17: Preliminary Science Report* (1973), states clearly that with the conclusion of the Apollo 17 mission—the last manned trip to the Moon—sufficient data and evidence had been accumulated to indicate that volcanic activity in the last three billion years on the Moon is either "highly restricted or virtually non-existent." Scientists today generally agree the Moon has been volcanically dead for the last 3 billion years!

What then were these lights seen on the Moon by the thousands over the last several centuries, which so many scientists attributed to volcanic eruptions? For instance, when the well-known astronomer Grover spotted a bright light on the Moon which lasted for 30 minutes, astronomers passed it off as another volcanic eruption. Again in 1958, when Soviet scientist Nikolai Kozyrev of the Crimean Astrophysical Observatory reported that he had spotted a bright "cloud" on or near the central peak of Alphonsus, it was passed off as "volcanic activity" by many scientists, although Kozyrev himself attributed it to fluorescing gases issuing from the crater's central peak.

On the night of November 3, 1958, Kozyrev photographed the spectrum of a reddish patch near the same place. The reddish light "seemed to move and disappeared after an hour." Strangely, it was passed off by most astronomers as another volcanic event—strangely because volcanoes are not generally known to move about.

At the same time that learned scientists claimed that these lights were actually fires of volcanic eruptions, they also claimed that the Moon was an airless world. And lights of fires—even glows from them—are impossible on an airless world. We know that the Moon is without an atmosphere. In fact, Dr. John H. Hoffman of the University of Texas points out: "If you took all the molecules in a cubic centimeter of the Moon's atmosphere and lined them up end to end, they would fit on the tip of your pen. But if you did the same thing with the air you breathe, the chain of molecules would reach to the Moon and back with some left over."

Admittedly, other scientists have passed off these bright lights and glows seen on the Moon as merely gases released from the lunar interior and fluorescing in the light of the Sun. Still others claim them to be nothing but solar radiation inducing ionization and exciting fluorescence.

But former NASA researcher William Corliss points out in his coverage that this does not seem to adequately explain the phenomenon either, since some have been reported on the dark or far side of the Moon. In fact, Corliss maintains that though many explanations have been offered (such as proton or thermoluminescence and, even less likely, glowing lava or fire-fountains of erupting volcanoes), "all have serious shortcomings as explanations of the phenomenon under consideration." (*Strange Universe.*)

What then could they be? Some could be solved without a doubt by such explanations.

Undoubtedly some of those huge glows seen on the Moon may be explained by some kind of thermoluminescence, as solar radiation fluorescing on the surface of the Moon, but certainly not all can be attributed to this. What are they then?

Could they not be UFOs? Interestingly, we do know that UFOs were seen by our astronauts on their journeys to and around the Moon. But our astronauts were not the only human beings to get a close look at such unidentified flying lights on or around the Moon—at least not if you believe Dr. Oscar Carter, a self-styled optics expert and amateur astronomer. Before the reader conjures up an image of a modern-day lunar Percival Lowell working at some giant observatory, it should be pointed out that Carter works with standard amateur equipment, although he claims he uses them with his own special optical inventions.

Carter, an investigator for the International UFO Registry, a worldwide Unidentified Flying Objects organization, was cited in that organization's journal *UFOlogy* as "a reliable investigator" who "believes UFOs are using the Moon as a base." (*UFOlogy,* Spring 1976.)

According to *UFOlogy*'s editor, Dr. D. William Hauck, Carter "has seen and photographed UFOs traveling near the Moon with the specially designed telescope of his own construction," which he "claims enables him to observe the area around the Moon with a clarity never before achieved." (*UFOlogy,* Spring 1976.)

However, Hauck does note that "astronomers at Mount

Palomar and the Naval Observatory have denied the reliability of Dr. Carter's sightings, the majority of which involved small black objects criss-crossing and reversing directions over the Mare Crisium and Oceanus Procellarum areas."

If there is any validity to Carter's sightings, they might have great significance. For, interestingly, these are the very same areas—the Sea of Crisis and the Ocean of Storms—we cited in our opening chapter as regions where strange, artificial constructions have been observed and photographed by both American and Soviet space probes.

However, that is the question—are Carter's sightings authentic and genuine? Whether or not there is any validity to them, there is a professional astronomer and astrophysical expert, Morris Jessup, whose credentials are impeccable, and whose reputation in the field put him in the forefront of discoverers of double stars, who is convinced lunar lights are UFOs. And he cites innumerable astronomers, many of them leaders in their fields, who claim to have seen unidentified moving objects on the Moon.

The professional judgment of this open-minded scientist, a leading astronomer and astrophysicist, after lengthy study of the strange things happening on our Moon, led to the unalterable conclusion that the strange lights and lunar "happenings" could be attributed to nothing else but Unidentified Flying Objects—spaceships on the Moon!

In the fifties—that incredible decade which brought a sudden rash of UFO activity to our Earth—this astronomer suddenly became interested in what UFOs could be. With his scientific background and keen mind, Jessup came to realize that flying saucers did exist and were undoubtedly operated by intelligences that were not of this world. Being well versed in astronomy, he soon came to recognize their base of operations—the Moon. For he knew that the many strange lights seen on the Moon and the mystifying "happenings" that have been reported as taking place on this mystery world could be attributed to nothing less than Unidentified Flying Objects. Jessup made an exhaustive study of the strange lights, glows, and clouds, the many moving objects, changes, and even disappearances of certain lunar features, and the strange appearance of so-called structures on the Moon that had been reported in scientific journals by leading astronomers all over the world. Jessup was aware that something was happening on our neighboring world, and after much study came to the shocking conclusion that in fact our Moon was inhabited—that it was undoubtedly the base of UFOs visiting the planet Earth.

Jessup was no crackpot. He brought to this controversial study an extensive background that had helped him build an impressive list of accomplishments. Besides teaching astronomy and mathematics at Drake University and the University of Michigan, he erected and operated the largest refractory telescope in the Southern Hemisphere for the University of Michigan, discovering numerous double stars which are now catalogued by the Royal Astronomical Society. His books on these astronomic discoveries and his cataloguing of various stars are in the University of Michigan library today, but not a single one of his controversial UFO books can be found there in their collection of flying-saucer books. That is what you call academic open-mindedness.*

In two books, *The Case for the UFO* (Citadel Press, 1955) and especially *The Expanding Case for the UFO* (Citadel Press, 1957), Jessup details the evidence that led him to conclude that not only are UFOs real and intelligently operated spacecraft but telescopic observation over three and a half centuries documented the fact that they have been on the Moon.

Many of the more impressive sightings that Jessup found in old astronomic records are listed in the NASA study. Some we have already analyzed in this chapter. Yet there are many others that Jessup came up with that are worth looking at.

In *The Expanding Case for the UFO* Jessup has a section entitled "Let There Be Light–on the Moon?" in which he presents this startling conclusion: "No single indication of UFO activity on the Moon is more intriguing than the unexplained intermittent lights."

He points out that a close study of these indicates that

---

*A NASA official charged that astronomer Jessup "never published a scientific paper in his life." A simple check at the University of Michigan graduate library reveals such important books as *New Southern Double Stars*, 1933.

they exhibit evidence of intelligent control: They sometimes fluctuate, "unlike the steady glare of reflected sunlight; sometimes they appear suddenly, shine for a few minutes or hours, and as suddenly disappear."

The sightings he cites are too numerous to detail here. But a few of the more impressive ones are worth looking at:

• A "speck of light," "very distinctly seen like a considerable star," sighted at the foot of the lunar Alps. The German astronomer Schroeter picked up a similar object here, although the British astronomer Birt does not consider them identical." (*Astronomical Register*, 1865, Vol. III, page 189.)

• Astronomer A. Fauchier of the Marseilles Observatory "was startled to see two bright points of light on the Moon." (*L'Astronomie*, Vol. VI, p. 312.)

• "A self-luminous spot on the Moon" seen in the dark body of the Moon, varying in intensity like "an intermittent light." (*Monthly Notices of the Royal Astronomical Society*, 1948, Vol. VIII, p. 55.)

• May 8, 1881—The astronomer Williams records nearly 1000 bright streaks and "light spots" in the Sea of Crisis, the very same spot where astronomers in the fifties reported seeing a huge bridgelike structure suddenly appear. On March 26, 1882, the entire area of the Sea of Crisis was "one mass of lights, streaks and spots." Other astronomers like Madler noted bright spots in the Mare Crisium area in May and October 1865. They appeared, moved about, and disappeared. Jessup concludes: "Such evanescent spots are puzzling unless we assume controlled activity of some kind."

• March 21, 1877—Mr. C. Barrett saw "the interior of the crater Proculus entirely lighted." (*The English Mechanic*, Vol. 25, p. 89.) This same volume describes an unusual sighting by Frank Dennett of a "point of light shining out of the darkness which filled the crater Bessel" and "a bright conical peak, surrounded by a circle of flat needle points clustering close to it." (April 19, 1877.)

And on and on go the myriad Moon lights. Jessup, an expert astronomer able to distinguish natural features from artificial objects, notes that some of these sightings indicate intelligent control. For instance, sometimes the lights exhibit "a brilliant glow . . . accompanied by flashes of singular brightness." Often it fades, often it moves about, sometimes appearing and reappearing. "Clearly, an indication that some intelligently operated mechanism was producing these lights," concludes Jessup.

Also, Jessup notes that these lights are often seen in certain key areas of the Moon—at least key in the light of where our astronauts went. Jessup observes that many sightings took place in the Sea of Tranquility, where our first astronauts were sent. "Well documented reports that Mare Tranquillitas at times of the full moon is sometimes covered with small bright specks of points."

We wonder, did knowledge of this influence NASA's decision to select this area for our first exploratory efforts? Remember, this is also the area where strange artificial-looking obelisklike structures were photographed.

The sightings of strange lights and lunar "events" that Jessup recounts come from the most experienced and expert professional astronomers as well as amateurs. The University of Michigan astronomer points out that many sightings of Sir William Herschel, whom he calls "the unimpeachable authority," note that he saw "some 150 very luminous spots scattered over the surface of the Moon during the total eclipse of October 22, 1790," alone.

But there are more definite and dramatic sightings which Jessup cites to prove this UFO thesis. Perhaps the most shocking is the sighting reported by the well-known astronomer Schroeter (1760). While watching the crater Cleomedes one night, Schroeter was surprised to see what appeared to be "a swirl of dust or vapor and a crater formed before his telescope."

Jessup believes that what Schroeter was witnessing was definitely the landing or taking off of a UFO!

But is such a radical explanation necessary? Could not this have been a volcanic eruption that Schroeter was watching? Some scientists thought it was. But, notes Jessup: "Bear in mind that in Schroeter's day it was almost universally thought that all lunar craters were of volcanic origin; meteorites could not yet be accepted as objects arriving from space much less a UFO. Schroeter could be pardoned for not assuming that he had seen a meteor strike the Moon. Yes, the 'eddying' must have been the dust created, agitated and blown aloft by the meteoric impact, or by the sudden movement, of or taking off, of a UFO." Since meteors of that size are extremely rare, and from other details of the sighting, it is clear that Jessup considers it to be a UFO. (*The Expanding Case for the UFO.*)

This was not the only instance that Jessup believes an astronomer might have seen a UFO take off. On November 20, 1878, the American astronomer Hammes saw an "uprush of something" which he reported to the U.S. Naval Observatory. He watched the whole thing for a half an hour. Asks Jessup: "Was it indeed, a landing or a blast off?" (*The Expanding Case for the UFO.*)

PLATO ONE OF THE CENTERS OF LUNAR UFOS?

One more area of the Moon that Jessup thinks might be the center of lunar UFOs is Plato, which he refers to as "the essence of the problem of life, intelligence and UFOs on the Moon."

Jessup catalogues many of the sightings in this area. Literally thousands of strange lights and happenings here alone indicate the Moon is occupied. In April 1781 no fewer than 1600 observations of strange lights were made here alone.* These lights often appeared in groups, sometimes even in what astronomers described as "geometrical arrangements." Many astronomers were puzzled by them, unable to offer the faintest lucid explanation. Jessup claims it is clear what they are—UFOs.

Jessup continued to examine the vast library of lunar "events" that clearly indicate life and activity on our companion world. They are too numerous to even mention in this already long chapter. Later we shall consider

---

* Birt, an English astronomer, deposited at the Royal Astronomical Society accounts of over 1500 sightings of lights, moving objects, and changes in light intensity on and around the crater Plato. Most of these took place in what Jessup calls "the Incredible Decade" of the 1870s.

some of the more startling "happenings" and developments on the Moon which the astronomer Jessup clearly considers evidence that the Moon is occupied in our present time. Suffice for now to give you Jessup's startling conclusion concerning our satellite:

> Reports of lunar activity which lay buried in the archives of the nineteenth century astronomical literature are now vibrating with new meaning ... I believe the discovery of life and intelligence in the environment of the earth-moon binary system is of as great ultimate importance to man as photographing the new galaxies, millions of light years away. I believe that the discovery, and our consequent awareness of this space intelligence, is of vastly greater and more immediate importance to us. It has the effect of putting us into a new world. Once this new world is established contemporary science will doubtless forget its opposition and claim credit for a new intellectual outlook.

Little did Jessup, as great an astronomer as he was, realize (at least to our knowledge) what a fantastic new world lay hidden within the Moon. But as we shall see—and this is the major thrust of this work—the overwhelming evidence is that spaceships from beyond our planet and probably from beyond our solar system are not just using our Moon as a base of operations, but the Moon itself appears to be a vast hollow spaceship!

Some experts believe that in a few years perhaps many lights will be seen on the Moon as man begins to develop and colonize his new lunar world. Arthur Clarke observes in his book *The Promise of Space* (Harper & Row, 1968): "It is strange to think that in a few more years any amateur astronomer with a good telescope will be able to see the lights of the first expeditions, shining where no stars could ever be, within the arms of the crescent Moon."

We have seen that they are already there—and have been for several centuries! Through our Apollo flights, Clarke's prediction in a sense has come to pass. The question we should now ask is: *Was man alone when he was on the Moon?*

The matter is urgent. They're no joking matter. They may well be ships from outer space.
   —*Dr. James McDonald*
      Professor of Meteorology, University of Arizona

THREE
THE LUNAR UFO CONNECTION

In the past few decades—and some researchers insist for many centuries—strange flying objects have been sighted in the skies of Earth. Unidentified Flying Objects have been reported over practically every country and continent of our planet—and in practically every era of history. Unquestionably, they have been seen in greater numbers in the last few years, and have created a controversy that has often led to heated debate not only among scientists but among people in every walk of life.

Since the 1950s brought flurries of Unidentified Flying Objects, many skeptics who once were sure that flying saucers did *not* exist have suddenly changed camps—perhaps none more dramatically than Dr. J. Allen Hynek, head of Northwestern University's Astronomy Department and of the Center for UFO Studies.

When Dr. Hynek (then director of Ohio State's Mc-Millan Observatory) was first asked by the U.S. Air Force to help them investigate UFOs, he was a complete skeptic:

"Oh, hell, that's all bunk," he remembers telling the Air Force officers who approached him.

But now Dr. Hynek has done a complete switch. Hynek the skeptic has become Hynek the believer. He has changed his mind about UFOs.

Hynek sums it up this way: "I challenge anyone to explain all the UFO reports in a rational manner."

Hynek points out that though most of the reports that something strange has been seen in the skies of Earth turn out upon investigation to be mistakes, "about 20 per cent of the reports are unexplainable." He insists that "about one in every five . . . come from highly credible sources and do not submit to rational explanations in our framework."

Hynek points out that, contrary to the popular notion that unexplainable sightings are few and far between, the dean of UFO researchers, addressing the Mutual UFO Network (a solid UFO group made up of scientists, college professors, educators, engineers, and people from many highly respected walks of life), told them:

"A paradoxical situation exists in the whole UFO problem area: we have too many sightings, not too few. Yet we are far from a solution. We are, frankly, embarrassed by our riches."

Hundreds of solid UFO reports come in every year—as many as half a dozen or more a day. And that presents a very difficult problem.

For the haunting question of our time is, where do all these Unidentified Flying Objects come from? Most investigators and UFO enthusiasts are convinced that they are really spacecraft carrying visitors from another planet (or planets) outside our solar system.

"That's the easiest interpretation," admits Hynek, "one that is in keeping with our level of technological development. But there are too damned many reports—there's an embarrassment of riches."

Hynek claims "there is an even deeper and more sinister embarassment of our riches, and that is what they imply about the origin of UFOs. While I, at least, do not feel quite ready to theorize about the ultimate origins of UFOs, the implication of the great number of reports per year is quite clear, and any theory of UFOs will have to explain their abundance. To our earth-bound minds, one or two Apollo missions per year is something we can understand; two or three Apollo missions per day would be quite another thing! Consider too that the nearest star to us is more than one hundred million times farther than the moon —well, I hardly need explain further!" (*MUFON Symposium*, 1973, p. 63.)

Admittedly, the problem of the vast dimensions of space

is just too great to allow conservative, orthodox scientists to think that so many spaceships (if they are that) are traveling here from faraway planets and star systems.

Dr. Carl Sagan, who thinks along these lines, gave the classic objection to UFOs on a CBS television program. Sagan, an astronomer from Cornell University who, ironically, believes that extraterrestrial spaceships may have landed on Earth thousands of years ago (as many as ten thousand times in our long history, speculates Sagan), nevertheless denies that there could be any visitors from space in modern skies. Sagan summed up his case with this tricky analogy in a radio interview:

"If you would believe, as the flying saucer cultists would have us believe, that the majority of saucer reports are due to their [extraterrestrial] visitations, then you have a very strange situation. That means that several spaceships are coming to the Earth over interstellar distances every day, as if all the anthropologists in the world were to converge on one of the Andaman Islands in the Indian Ocean because they had just invented the fish net or something." (Daniel Cohen, *Myths of the Space Age*.)

Admittedly, it is hard to conceive how spacecraft by the hundreds and even the thousands are coming to our planet from other far-flung star systems. Theories abound that they may be from bases inside our Earth, even under the oceans. A few even believe they may be coming from Mars.

There is another area which was once considered by many major UFO researchers as the most likely base for UFOs—and this is our Moon. Even Dr. Carl Sagan in his younger and less conservative days theorized that UFOs could possibly be based on the Moon—on the far side, which Earthlings up to that time had never seen.

UFOS BASED INSIDE THE MOON?

Another UFO researcher, Major Donald Keyhoe, whose recent book *Aliens from Space* (Doubleday, 1973) has again placed him among our foremost UFO writers and researchers, once was quite convinced that extraterrestrial astronauts came from our Moon. He asserted in his book

The Flying Saucer Conspiracy: "All the evidence suggested not only the existence of a Moon base, but that operations by an intelligent race have already begun. If so, who could the creatures be? Were they from other planets or did they originate on the Moon?"

It was with anxious expectation then that many UFO enthusiasts watched our Apollo flights to the Moon. Various verified reports had come back that our Mercury and Gemini astronauts (as well as various Soviet astronauts) had seen UFOs around the Earth on their journeys into outer space. In fact, Dr. Garry Henderson, one of America's top research scientists, claims that "all our astronauts have seen these objects [UFOs] but have been ordered not to discuss their sightings with anyone." He also maintains that NASA has actual photos of these craft, taken at close range by still and movie cameras."

We do know that NASA has admitted that a few astronauts have indeed seen and even photographed unexplainable objects in outer space. The notorious *Condon Report: Scientific Study of Unidentified Flying Objects* (Bantam, 1969), which attempted to whitewash all UFOs out of the skies of Earth, considered and weighed reports of UFO sightings on Gemini 4, 7, 9, and 11. Interestingly, Chapter 6 of that study includes detailed descriptions of these four separate sightings by orbiting astronauts, which the Condon report was unable to explain, except perhaps the sighting of Gemini 11. (See *Our Mysterious Spaceship Moon.*)

That chapter was done by Dr. Franklin Roach and was one of the few areas of open-minded consideration given to the entire UFO problem researched by the study.

Stanton Friedman, nuclear physicist and space scientist now devoting full time to the problem of UFOs, in his book *UFOs—Myth and Mystery* states: "This chapter by Dr. Franklin Roach is one of the better chapters of the Condon report, though Condon's summary fails to give adequate attention to these observations. In fact, frankly they are dismissed as 'unexplainable.' "

THE QUESTIONABLE APPROACH OF THE CONDON REPORT

Of course, this is not to say that the *Condon Report* allows even the possibility that flying saucers exist. This study, sponsored by the U.S. Government and Air Force and approved by the National Academy of Sciences, is condemned by Dr. J. Allen Hynek, an astronomical consultant on Project Blue Book, the official Air Force investigation of UFO reports. Twenty years of experience gradually turned this highly respected investigator from a skeptic into a believer in the reality of UFOs.

Hynek objected with stinging criticism to the unscientific approach of the *Condon Report*. He became convinced that UFOs did exist, although what they were he admitted he could not say. Dr. Bruce C. Murray, expert in planetary sciences at California Institute of Technology, reviewing Hynek's work *The UFO Experience* (Henry Regnery, 1972), agrees that the *Condon Report* seems to reflect more a "desire to make the UFO problem vanish altogether from scientific jurisdiction than a thoughtful attempt to isolate possible genuinely new empirical observations."

He adds: "Thus Hynek not only defends UFO's but necessarily attacks the scientific establishment that has written them off. Can our modern scientific institutions be as limited as their predecessors were when scientific authority refused to acknowledge the reality of meteorites, hypnosis, continental drift, germs, Troy, Atlantis, and Pleistocene Man?" (*Science*, August 25, 1972.)

As head of the Center for UFO Studies, a clearinghouse for UFO reports from all over the world, Dr. Hynek has collected thousands of reports not only of strange lights in the skies but of objects seen on radar as well as by competent witnesses, and, even more amazingly, of encounters with spacecraft and their occupants at close range (within 20 to 500 feet) and with "humanoid figures reportedly observed as well." Hynek calls these "close encounters of the third kind." Yet Dr. Murray, who is an objective reporter, criticizes Hynek despite all his expertise, for even the leading UFO scientist in the country "obviously feels uncomfortable" about including cases of Close-Encounters of the Third Kind* because of what Murray calls "the little green men" implications. Yet Murray maintains correctly these should not be ruled out just because they are bizarre.

Still Hynek deserves a world of credit for doing just that. He has become so impressed over the years with the evidence that he makes a plea that scientists "set aside [UFOs] bizarre aspects" and investigate them objectively, "for sufficient scientific respectability for the UFO subject to permit modest federal research funds to be awarded to it and new data to be gathered without fear of ridicule." (*The UFO Experience.*)

This is the very same plea we make in regard to the mystery world that circles our Earth—that serious consideration not only be given to the possibility that it is the source of many of the UFOs visiting Earth skies but that a serious examination of the evidence that it may be a huge, hollowed-out orb converted from a natural planetoid into a kind of spacecraft and "driven" from an unknown region of the Universe into orbit around our planet, be undertaken.

Admittedly, this theory is so bizarre that the natural reaction may be to at first consider it preposterous. But it is possible even though our own earthbound minds might find it at first not plausible or even feasible. We ourselves reacted this way and it was not until the overwhelming mountain of evidence broke our shell of skepticism that we began to consider it seriously. We feel along with Hynek that this theory too should be given serious scrutiny by the scientific community who should undertake an analysis of the data and evidence. Because if it is true that our Moon is really a spacecraft—a hollow world—then this has got to be the most important discovery in the history of humanity. Its implications are staggering.

For it is not enough to determine whether or not UFOs

---

* Close Encounters of the First Kind are those UFO sightings made at close range (less than five hundred feet) where there is no contact or interaction of any kind.
Close Encounters of the Second Kind occur when a UFO is not only sighted at close range but where it leaves a visible record of its visit, such as chemical changes in the ground over which it hovered or on which it landed.

really exist, but also to determine their source or at least their nearest base of operation. As we have seen and shall continue to see throughout this book the evidence seems to point to the Moon as their point of origin. Indeed, our astronauts on their way to the Moon and while on the Moon encountered UFOs. Such reports can be verified through the logs of astronaut-Mission control conversations.

Some additional reports come from alleged accounts, which are unfortunately not verified. Let us first take a look at a few of these.

APOLLO 8 SPOTS HUGE OBJECT ON THE MOON?

In addition to the "unexplainable" flying objects sighted by our Gemini astronauts, which even the government-sponsored *Condon Report* could not reasonably explain, a number of other strange reports have come to light. The first trip to the Moon was made by Apollo 8 astronauts Frank Borman, James Lovell, and William Anders in December 1968. As the Apollo crew approached the Moon (December 24, 1968), where they were to begin checking out prospective landing sites, they ran into something unexpected. After going into the orbit and traveling to the far side of the Moon, the Apollo 8 astronauts sighted a huge extraterrestrial object of some kind, which they managed to photograph. They estimated the size of this unusual object—whatever it was—to be about 10 square miles!

On the next orbit they peered all around with anxious eyes, trying to spot it again. But, strangely, the huge object had disappeared. Where could such a vast thing have gone so suddenly? What is even stranger, subsequent photos of the site where they saw it showed not a shred of evidence that anything had landed there recently. Could it have disappeared inside the Moon, to an underground base in a hollow portion of the lunar orb? No one knows.

Speculation ran rampant as to what it could have been: a lunar space station of some kind set up by alien beings from inside the Moon? Or was it a UFO vehicle of some kind—an extraterrestrial spacecraft merely checking out our Apollo mission to the Moon?

To this day no one has given an adequate explanation of this alleged sighting. And no one knows where the object came from or disappeared to. It remains totally unexplained. An absolute mystery.

MYSTERY SPACECRAFT VISITS APOLLO 10

The next trip back to the Moon was Apollo 10, and here again UFOs made their appearance. On this mission astronauts Eugene Cernan, Thomas Stafford, and John Young were to try out the lunar landing vehicle in lunar skies for the first time. They were to duplicate every maneuver that the first manned landing mission to the Moon would carry out, except, of course, the actual landing itself. Their object was, as one astronaut put it, to "snoop" around the Moon, trying out the Lunar Excursion Module (or LEM) and at the same time looking over four prospective landing sites for the Apollo 11 landing mission. Because of this, the astronauts dubbed the lunar landing module *Snoopy*, the command module they naturally nicknamed *Charlie Brown*.

Charlie Brown went into orbit around the Moon on May 22, and *Snoopy* soon was flying for the first time in lunar skies, descending to within 50,000 feet of the Moon's surface, the closest any humans up to that time had been to the Moon. As *Snoopy* dipped to within 4.5 miles of the Moon's surface, suddenly a UFO rose vertically directly from below to greet them. The astronauts of Apollo 10 allegedly caught a glimpse of this brief encounter with a lunar UFO and, moreover, captured it on 16-mm motion picture film. They also took a number of still photos of this unpublicized sighting.

APOLLO 11's RENDEVOUS WITH UFOS

In our book *Our Mysterious Spaceship Moon* we cataloged in two chapters various sightings of UFOs from Project Mercury and Gemini missions through the Apollo flights. More importantly, we published official transcripts taken from NASA Mission Control–astronaut conversations covering the various sightings of UFOs. These included the dramatic sightings made by astronauts Neil Armstrong and Edwin Aldrin, Jr., the first men on the Moon, along with their trip mate Michael Collins, command module pilot.

En route to the Moon on their very first day in space, the crew of the *Columbia* (the Apollo 11 command module) sighted a strange object hovering high above Earth. They managed to take films of this "bogey," as the astronauts termed Unidentified Flying Objects. In our first book on moon mysteries we included a debriefing-session made with the Apollo 11 astronauts, who discussed this strange sighting. Yet to this day NASA officially insists that no Apollo astronaut ever sighted a UFO.

APOLLO 11 SIGHTS UFOS OVER THE MOON

This was just the beginning of an entire series of strange sightings for Apollo 11.

At 6:00 P.M. (Eastern Daylight Time on Earth) on July 19, 1969—the day before the history-making lunar landing—astronaut Edwin Aldrin, Jr., who had just entered the *Eagle* (lunar landing craft) for final testing of all the craft's systems, was doing some last-minute filming of the Moon's surface with a 16-mm motion picture camera, checking for possible landing hazards and giving scientists and space officials back on Earth additional lunar-surface information, when suddenly *two* UFOs came into view.

One appeared to be larger and brighter than the other. Both were traveling in vertical position in respect to the Apollo spacecraft, which now was in orbit around the Moon. They were moving at a fairly fast clip, both objects moving now horizontally to the center of the camera's view.

Suddenly, both "bogeys" reversed their direction, retracing their path through the Moon's skies until they disappeared to the left of the astronauts' field of vision.

Seconds later, the mysterious objects reappeared. This time they were flying above the Apollo 11 spacecraft, and began slowly descending toward it. Aldrin turned his camera 90 degrees, to catch them in his lenses, and, strangely, the two UFOs seemed to hover and come to halt for a moment—almost as if they were permitting the astronauts to film them.

Buzz Aldrin, veteran of the Gemini 12 mission, during which he sighted four UFOs, ground away with his camera, taking invaluable (but now secret) footage of the two mysterious objects. The two brilliant UFOs continued to descend, and suddenly Aldrin noticed a brilliant emission extending between the two craft. Robert D. Barry of 20th Century UFO Bureau, who commented on these remarkable sightings, says: "Speculation at the time was that this 'trail' was possibly connected to the vehicles' motational systems, possibly even exhaust."

As the astronauts watched, dumbfounded, the two UFOs, which seemingly had joined together, suddenly separated and began to rise vertically at a fast rate, finally disappearing from sight. The astronauts reported that upon separation, each "bogey" or spacecraft became somewhat brighter and seemed "to emit a force field taking the form of a blurry 'halo' around the entire craft." (*Modern People Press UFO Report*, p. 9.)

This was not the end of the strange sightings, since a little while later one of these circular craft came back into camera view, only to quickly disappear from lens range. It is reported that "during this sighting, 10 other egg-shaped objects were seen flying in the foreground of the camera's view."

As Robert Barry puts it: "Naturally, NASA did not release these photos to the general public, taking great pains to edit any such mysterious craft from the final stills which were released to major newspapers and magazines around the country." (*Modern People Press UFO Report*.)

But this enterprising researcher managed to get them and make them public, thereby doing a service to everyone— for the American taxpayers, who footed the entire bill for the multi-billion-dollar Apollo flights, deserve no less.

These are the major sightings made by our Apollo 11 astronauts. In addition, there are two unauthenticated reports of sightings by this same astronaut team as they made their way to and from the Moon. One was reported by Otto Binder, former NASA researcher and writer, who claimed that certain sources with their own VHF receiving facilities that bypassed NASA broadcast outlets overheard a startling astronaut conversation. Apparently, as Aldrin and Armstrong were making their rounds on the Moon, collecting lunar samples, Armstrong exclaimed, "What was it? What the hell was it? That's all I want to know."

Mission Control, alarmed at what was taking place, responded: "What's there? . . . (garble) Mission Control calling Apollo 11 . . . Apollo 11: These babies were huge, sir . . . enormous . . . Oh, God you wouldn't believe it! I'm telling you there are other spacecraft out there . . . lined up on the far side of the crater edge . . . they're on the moon watching us . . ." (*Saga's UFO Special,* III.)

Binder says NASA officials made sure that these words never reached the public. The broadcast, on a five-second delay, like all Apollo broadcasts, was censored.

The other *unauthenticated* report of a startling UFO sighting by our Apollo 11 astronauts on the Moon comes from *Modern People Press UFO Report,* which in an article entitled "Unexplained Lunar Mysteries Point to Intelligent 'Moon Men,'" said that a national newspaper carried the sensational story that as the crew approached the Moon for a landing they could see a "space fleet" lined up on the surface of the Moon! This sighting has never been verified, but we give it to you for what it is worth:

"APOLLO 11 ASTRONAUTS: 'You can see them lined up in ranks on the crater's edge! It looks as if someone got here before us.'

"Mission Control then broke in and ordered the astronauts to film the objects. Those films have never been released and the story itself has never been confirmed."

Here is a brief summary of the highlights of the other UFO sightings made by our various astronauts:

- Apollo 12—Sighted three UFOs halfway to the Moon. Again the astronauts reported hearing various noises—whistles, fire-engine sounds, beeps—interfering with their communications to Mission Control. On the return trip the astronauts reported seeing another UFO just before splashdown in the Pacific (over Burma).

- Apollo 15—A mysterious "flying object" flashed across the lunar skies briefly, just missing, it is claimed, astronauts David Scott and James Irwin.

- Apollo 16—Astronaut Thomas Mattingly, orbiting in the command module above the Moon, saw a "light" flashing across the Moon's skies. It disappeared in a few seconds beyond the lunar horizon. Dr. Farouk El Baz, former NASA scientist, claims that such lights as this seen by our astronauts "must remain in the category of UFOs. . . . It is certain [they] were not any spacecraft we know of because it was moving too fast. No Russian or American spacecraft can move that fast, either on or near the Moon." (*Saga,* April 1974.)

- Apollo 17—Two more UFOs sighted by our astronauts, this time by Ron Evans and Harrison Schmitt, the only full-fledged scientist sent to the Moon.

This summary of our many UFO sightings made around the Moon by our astronauts indicates that nearly every Apollo astronaut and mission encountered UFOs on or around the Moon. Why didn't the public hear of such sightings during the well-publicized and media-carried Apollo trips? No on-the-spot reports came through. Then why was there such a flurry of alleged reports later?

Stanton Friedman, who spent many years working for the U.S. Space Agency, points out: "It should be noted that the Apollo flights, about which there have been rumors concerning UFOs, employ a different approach to communications between the astronauts and the ground. The radio signals are sent directly back to Houston and then rebroadcast with Houston having the option of deleting whatever they choose to delete with essentially no one outside NASA able to monitor the broadcasts. During the earlier Gemini and Mercury flights the talk was 'in the open'—readily monitored by ham equipment." (*UFOs— Myth and Mystery.*)

Also, an additional censoring system was the use of a code system and private channels to which the astronauts and Mission Control could switch at a moment's notice. In our book *Our Mysterious Spaceship Moon* we included actual Mission Control–astronaut conversations of the sightings of UFOs. In these transcripts Mission Control from time to time gave orders to the astronauts in some unusual terms, seemingly whenever an unusual sighting was taking place, directing them to "Go to Whiskey Whiskey" or "Barbara Barbara" or "Bravo Bravo" or "Kilo Kilo."

One NASA scientist who has left the space agency admitted that private channels were set up by NASA and used during these missions and that these strange terms could have been code names for switching to such channels.

Interestingly, there is an intercontinental ballistic missile base in Montana by the name of Kilo Kilo. Could NASA have used this base's radio equipment to filter out through this prearranged private channel anything NASA did not want the public to know? Shockingly, Whiskey Whiskey, Barbara Barbara, and Bravo Bravo are also bases in the West.

To this day NASA has not revealed officially that our Apollo astronauts did actually see UFOs while in outer space. Stanton Friedman, widely known scientist who now works full-time investigating UFOs, observes: "I did once spend over two hours discussing UFOs with one of the Apollo astronauts who was extremely interested in what I had to say, bought four hard-to-get UFO documents but would give me no information at all—not even that which turned up in the Condon report—about astronaut sightings."

Friedman, longtime space scientist himself, then points out why our astronauts have remained so mum:

> For those who have never worked on classified programs I should stress that the classification guides for such programs are normally also classified so it is, practically speaking, impossible to even determine what fraction of the astronauts' observations of the earth and space is classified. Contrary to some comments I have heard much NASA data is secret. The penalties for breaking security are extremely rigorous and it is often necessary to lie as a cover story. I believe I can guarantee as a result of working under security for 15 years that the government and its employees can indeed keep secrets—despite the fact that

some secrets have been inadvertently revealed. (*UFOs —Myth and Mystery,* p. 7.)

Although our astronauts have kept a tight lid on UFO secrets, it is interesting to note that several have openly expressed themselves privately on the subject, stating their conviction that UFOs do indeed exist.

- Astronaut John Young, the ninth man to set foot on the Moon, speaking on the existence of UFOs:
"If you bet against it you'd be betting against an almost sure thing . . ."

- Astronaut Edward Mitchell, the sixth man to set foot on the Moon, commenting on UFOs:
"The only question that remains is, where do they come from?"

- Astronaut Eugene Cernan:
"I believe UFOs belong to someone else and that they are from some other civilization."

CRACKS IN OUR SPACE AGENCY'S CURTAIN OF SECRECY

Despite all the revelations to the contrary, NASA officials still adamantly deny that our Apollo astronauts saw any UFOs in space. Gordon L. Harris, NASA official at the Kennedy Space Center, was quoted recently as stating that "no official information which would in any way substantiate the presence of UFOs in space—including the post-mission reports of the Apollo astronauts—is being withheld from the public." (*Modern People Press UFO Report,* p. 10.)

But Robert D. Barry of the 20th Century UFO Bureau points out that Donald L. Zyistra, chief of NASA's Public Information Branch in Washington, D.C., did make this astounding statement: "While NASA has no record of UFO sightings as such, during the Apollo manned-flight missions there were sightings from the spacecraft which our astronauts were unable to explain."

Flying objects which must be classified as "unexplainable" fit precisely the definition of UFOs—they are UFOs!

NASA PRIVATELY ADMITS THE EXISTENCE OF LUNAR UFOS

An even more startling revelation comes from Dr. James Harder, an engineering professor at the University of California, who told a university symposium that several Apollo flights were followed by UFOs. In a story carried by United Press Wire Service, Dr. Harder said he discovered the Apollo-UFO incidents while reviewing tape-recorded conversations between our lunar spacecraft and NASA's Houston space control.

According to Dr. Harder, the taped conversations (published in *Our Mysterious Spaceship Moon*) clearly show that the Apollo 11 spacecraft was followed halfway to the Moon and a UFO definitely trailed Apollo 12 on three orbits around the Moon. Dr. Harder confronted NASA authorities with these facts, and they shockingly admitted privately to him that "they had suppressed the UFO incidents for fear of public panic."

Harder also claims that a member of the Apollo 12 space team (whom he declines to identify) admitted to him that the mission had spotted UFOs.

The University of California professor points out that the official explanation—that the objects were part of the spacecraft trailing behind—was not supported by the speed observed on NASA Earth-based instruments.

This in general has become the commonplace explanation for all UFOs that were seen by astronauts in outer space: They were (the public is told) merely debris from the spacecraft or the rocket missile system.

And so it is not surprising that when astronaut Frank Borman was privately asked by an interested airline pilot if he had indeed seen UFOs in space, as reported, Borman denied that they were actually Unidentified Flying Objects. The ex-astronaut fell back on that old NASA song: They were fragments from the launching vehicle.

Yet even the *Condon Report,* which tries so hard to latch

on to any answer to explain away all UFOs, does candidly admit "this is impossible if they were travelling in a polar orbit as they appeared to the astronauts to be doing." Thus, even this government-backed study of unidentified flying objects rejects the NASA explanation and finally lists the UFO sighting of Gemini 7 as "unexplainable." (*Condon Report*, pp. 207–208.)

Interestingly, the photos that astronauts Borman and Lovell took of the mysterious objects show clearly they were not space debris. One picture, in fact, shows a bright, shining UFO, and another, surprisingly, three glowing objects.

THE FINAL PROOF OF THE GREAT NASA COVER-UP

The proof that there is actually a cover-up comes from Robert D. Barry of the 20th Century UFO Bureau, who claims that a classified NASA document (KMI-8610.4) that he managed to get hold of actually orders astronauts that "sightings of objects not related to space vehicles" (UFOS) had to be reported as pieces of regular NASA spacecraft.

This same source also insists that a top NASA official actually stated: "Every manned mission we have sent to the Moon has been under surveillance by UFOs." (*Modern People Press UFO Report*, p. 6.)

A study of the records and a glance at the photos will convince even the most diehard skeptic that this is exactly what happened when man went to the Moon.

UFOS BASED INSIDE THE MOON

One final question might be haunting you: How is it possible that so many sightings were seen on the Moon when no bases were spotted on the lunar exterior? While it is true that there are no *known* bases (some observers believe that several hundred strange domes that cover certain craters on the Moon *may* be just that), it is quite possible

56

that lunar bases exist *underground*. Admittedly, again this is only theory.

However, this theory—that the Moon might internally harbor an entire series of bases for UFOs—is not without evidence. As we have seen, the Moon certainly has been the center of a lot of UFO activity. Thousands of strange lights, "happenings," and even structural changes have been observed, as documented by leading astronomers and scientific experts the world over. Some scientists like H. P. Wilkins believe there is substantial evidence that the Moon might have extensive hollows. These, if they exist, obviously could harbor such alien beings and their bases.

In fact, according to a startling new theory of two Soviet scientists, the Moon may not only have hollow areas internally but itself be hollow! These two bold theorists have compiled some impressive (though not conclusive) evidence that our Moon has also an inner shell or "hull" of metal, and they insist that there is evidence of artificial construction existing inside the Moon! Mikhail Vasin and Alexander Shcherbakov's mind-blowing theory is the major theme of this book and shall be examined and scrutinized in great detail.

Strangely enough, according to George Leonard, an amateur astronomer and author of the mind-boggling book *Somebody Else Is On the Moon* (McKay, 1976), scientists on this side of the Iron Curtain are also arriving at the same conclusion—that our Moon is a spaceship! Leonard says that a former NASA scientist related this information to him.

This pseudonymous ex-NASA researcher, Dr. Sam Wittcomb, "heard it [the spaceship theory of the Moon] explained by an engineer at the Jet Propulsion Lab [NASA] and by a British physicist at Oxford. The theory is that the Moon is a vast spaceship, that it was driven to our solar system many thousands of years ago after suffering a terrible calamity in space. Its occupants have been engaged in a long, slow effort to repair the damage. Machinery is seen in several places on the Moon. It is nuclear-powered, and will one day be used to drive the Moon out of our orbit into space again."

The problem with such an unusual theory is that scientists as well as the general run of people are wholly unfamiliar with such a far-out technological concept.

Dr. Farouk El Baz, former NASA scientist who trained astronauts in geology, makes this profound observation: "We may be looking at artifacts from extraterrestrial visitors without recognizing them."

Could it be that the biggest such extraterrestrial artifact is the very strange world of the Moon itself?

If compelling evidence can prove that this satellite of ours is truly a spaceship, then there is such an extraterrestrial artifact in our corner of the cosmos; our Moon is then in reality a huge UFO that so far has not been recognized for what it is—and a UFO, incidentally, that everyone can raise his eyes to the heavens and see.

Furthermore, if it can be proven beyond a shadow of a doubt that our Moon is really an alien spacecraft—a huge, hollowed-out planetoid—it could be the key, as we have seen, to the streams of alien spacecraft that are coming to the planet Earth.

Then the lights of the Moon, which move around, come and go, appear and disappear, would make sense. As Dr. Maurice Jessup, who figured it out several decades ago after a thorough study of strange lights and "happenings" that took place on the Moon, concluded:

"It is no longer necessary to explain the visitors as coming from Mars, Venus or Alpha Centauri; they are a part of our own immediate family, a part of the earth-moon binary system. They didn't have to come all those millions of miles from anywhere. They've been here for thousands of years. . . ."

We shall see that compelling evidence can be marshaled to prove that they come from our Spaceship Moon.

I keep telling these geologists to let their imaginations go!
         —*Dr. Harold Urey, Dean of Moon Scientists*

FOUR
IS OUR MOON A KIND OF HOLLOWED-OUT SPACESHIP?

From time immemorial people have gazed into the heavens and wondered about the mysterious glowing world revolving around us. Man has marveled at the Moon's beauty, been mesmerized by her radiance, even made more amorous by this "silver ornament of the night," as one author back in the late sixteenth century described her. And even before history began, there is considerable evidence that men all over our globe worshipped her. Now, in modern times, Earthbound man has broken the chains of gravity, to set forth from our planet and set foot on this once-unreachable world.

Many scientists of Earth thought that going to the Moon would enable man to divine the secrets of the universe. To scientists this alien world—the first that man ever set foot upon—would prove to be a cosmic Rosetta Stone that would enable mankind to read the records of the cosmos and unravel the origins and nature not only of the Moon itself but also of our solar system.

However, instead of shedding light, this longed-for Holy Grail of the scientific world turned out to be a cornucopia of mysteries. As rock after rock was returned and voyage after voyage completed, the results, mulled over and studied, merely produced mystery after mystery, puzzle after puzzle. Instead of answers, only confusion has been spawned. In fact, so much conflicting and perplexing information was produced that after the second Apollo mission Dr. Louis Walter of NASA's Space Flight Center confessed: "In view of this latest data, these guys are really staying loose on their feet." No one wanted to commit himself. And that has been the whole story of the Apollo scientific findings so far.

Despite this, a few bold scientists are hesitantly advancing theories, none of which is being wholeheartedly accepted by colleagues. Scientists who for the most part have been scrambling over all this information and wrangling over its meaning do admit that most of the old theories do not jibe with already established facts. In truth, some of the answers often seem to be more complicating and contradictory than the now-shattered theories of the past.

The Moon itself remains a cosmic conundrum that scientists have just not been able to crack. Although the major theories have been shattered under the surprising evidence, the amazing outcome, as we view it, is that nearly all the information and data that we have pored over, nearly all NASA's findings, seems to support the rather bizarre Soviet theory that holds that our Moon is not entirely what it seems—not completely a natural world.

Two Soviet scientists, Mikhail Vasin and Alexander Shcherbakov, in a Soviet government publication a few years ago (July 1970) claimed that our Moon is hollow and may well indeed be a spaceship built by some unknown alien intelligence from a far-flung star system.*

A COSMIC NOAH'S ARK?

These bold scientists tell us:

> It is quite likely that our moon is a very ancient spaceship, with an interior that was filled with fuel for its engines. . . . The hollow interior should also still

*Although a NASA official claimed that the Vasin-Shcherbakov article was published in a government scientific journal as a "spoof," adding that "the hollow moon theory was entirely a joke," the author has received letters and material directly from Vasin, who has in fact published his latest proofs on the artificial moon theory in a book (1976), a copy (in Russian) of which he was good enough to send me, and which we received just as this book was going to press.

contain materials and equipment for repair work, navigation instruments, observation devices and all manner of machinery.

In other words, the huge spaceship carried everything necessary to serve as a kind of Noah's Ark of intelligent creatures on a voyage through the universe thousands of millions of years.

Who were these highly advanced creatures who produced in their mind a spaceship requiring a technology and vision we haven't yet even approached?

The two Soviet scientists refuse to speculate on this question. They nevertheless do state: "Perhaps it was even the home of a whole civilization of creatures whose original home planet could no longer sustain life."

Why would two highly respected, orthodox scientists from a highly respected science institute propound such a preposterous and, in their own words, even "crazy" theory? As they themselves confess:

"Abandoning the traditional paths of 'common sense,' we have plunged into what may at first sight seem to be unbridled and irresponsible fantasy. But the more minutely we go into all the information gathered by man about the moon, the more we are convinced that there is not a single fact to rule out our supposition. Not only that, but many things so far considered to be lunar enigmas are explainable in the light of this new hypothesis."

Here are the five major areas of evidence upon which these two Soviet theorists formulated and postulated their "crackpot" theory*:

(1) THE MYSTERY OF THE ORIGIN OF THE MOON

The origin of the Moon remains a puzzle, and all three major theories as to where the Moon came from and its origin must be rejected.

The Soviet scientists first of all examine these three major theories of the Moon's origin and reject them all:

HYPOTHESIS I. The Moon was once a part of the

* For the complete Soviet "Spaceship Moon" theory and the Soviet scientists' position paper, see *Our Mysterious Spaceship Moon*.

Earth and broke away from it.
This has been refuted by the evidence.

HYPOTHESIS II. The Moon was formed independently from the same cloud of dust and gas as the Earth, and immediately became the Earth's natural satellite.

The Soviet researchers reject this, too, not only because of the big difference in the specific gravity (density) of the two radically different worlds but because analysis of Moon rocks indicates that our companion world is "not of the same composition as the Earth's."

HYPOTHESIS III. The Moon came into being separately and, moreover, far from the Earth (perhaps even outside the solar system).

This is the so-called capture theory—that the Moon, wandering through our solar system, just happened, "by a complex interplay of forces . . . [to be] brought within a geocentric orbit, very close to circular. But a catch of this kind is virtually impossible, the Soviet scientists note."

Vasin and Shcherbakov conclude:

"In fact, scientists studying the origin of the Universe today have no acceptable theory to explain how the Earth-Moon system came into being."

This then leads to their own theory:

"Our hypothesis: The Moon is an artificial Earth satellite put into orbit around the Earth by some intelligent beings unknown to ourselves."

(2) SCIENTIFIC EVIDENCE THAT THE MOON MAY BE HOLLOW

The major thrust of their density argument is that density studies of the Moon indicate that the Moon might be hollow. If it is, then, as all astronomers agree, it would have to have been artificially hollowed out.

Vasin and Shcherbakov point to the "big difference between the specific gravity of the Moon (3.33 grammes per cubic centimeter) and that of the Earth (5.5 gr.). Such a low specific gravity indicates that the Moon could be hollow."

(3) THE EVIDENCE THAT THE MOON MAY HAVE AN INNER SHELL OR HULL OF METAL

The Soviet scientists insist there is evidence that the Moon's inner and outer shells were in part formed by the hand of alien intelligence—that they are in fact at least partly of alien "construction." According to their theory, the Moon has two such areas: the outer shell of rock and dirt, which they maintain shows clear evidence of reinforcement (the so-called seas or maria, which the next section will detail), and an inner shell or hull, which they estimate to be 20 miles thick. This inner hull served as space armor to protect the occupants of this huge spacecraft in their journeys through the universe.

(4) THE EVIDENCE THAT THE MOON SHOWS SIGNS OF HAVING BEEN HOLLOWED OUT AND REINFORCED (SPECIFICALLY, THE DARK CIRCULAR SEAS, THE SO-CALLED MARIA)

The Soviet researchers are convinced that the flat areas or plains regions of the Moon, called seas (maria) because once people on Earth thought they were actually covered with water, are in reality made up of metallic rock—evidence of an artificial outpouring from the Moon's interior. These dark, level areas, which can easily be seen from Earth on a clear night when the Moon is full, are actually loaded with dark mineral ores like titanium and iron, along with other rare metallic elements.

Three of these huge dark spots form the so-called Man in the Moon: the Sea of Rains, which makes up the right eye, the Sea of Serenity, and the Sea of Tranquility, the left eye.

The Soviet "spaceship" theorists point to the strange discovery of great amounts of metallic elements in the rock samples brought back from these lunar seas—such metallic elements as titanium, zirconium, chromium, and the highly refractory elements yttrium and beryllium—all of which are mechanically strong and resistant to high temperatures.

THE OUTER METALLIC HULL OF A SPACECRAFT?

Vasin and Shcherbakov point out:

"This is the perfect kind of material out of which to fashion and reinforce a spacecraft. Such metals were used not only in Spaceship Moon's inner hull but its outside exterior shell to withstand the rigors of their long space odyssey.

"The metals were chipped from the armor-plate and fused with the loosely packed dirt and rocks in the regions around where they hit by the super-hot meteorites to form rocks."

PROTECTIVE COVERING OF THE SPACESHIP MOON?

One reason Vasin and Shcherbakov suspect that the Moon has such inner and outer spaceship shells made out of strong metallic elements is simply that although massive meteors and even huge asteroids have crashed into the Moon by the millions—some of them but a few feet across and others over 100 miles in diameter—and although they struck like explosive missiles with the force of many tons of TNT, they merely dented the outer shell of rock and dirt. How come? The Soviet scientists point to a study done by a fellow Russian of the expected depth of Moon craters created by such fierce impact. Professor Stanyukovich's study concluded that these celestial missiles should have penetrated to a depth equal to four to five times their own diameter (24–30 miles at least). Yet almost invariably Moon craters, even those 100 miles or more across, are a mere mile or two deep at most.

Something underneath the Moon's layer of rock and dust stopped them when they began to penetrate. The metallic particles in lunar samples indicate this was a shell of metallic rock which the meteors and other heavenly missiles crashing into the Moon could not penetrate.

Another anomaly is that, despite their huge size, many lunar craters are convex—not concave, as we would expect them to be if huge meteors exploded into mere rock and dirt. Again, it was the tough protective shell that prevented this.

(5) THE MYSTERIOUS MASS CONCENTRATIONS (MASCONS) INSIDE THE MOON

Strange, unexplainable mass concentrations exist under these mystifying maria. These puzzling areas exist in the flat, circular plains regions and were discovered to exert a heavier or stronger pull of gravity than the other areas of the Moon.

The reason for this, claim Vasin and Shcherbakov, is simply that something must exist there at the bottom of these maria that causes this heavier pull: "the stocks and the cement-metallic like repair materials still stored beneath those regions and the many pieces of heavy repair equipment. The concentration of all this weight under the circular lunar 'seas' would be enough to make the pull of gravity greater," the Soviet scientists speculate.

A SCIENCE-FICTIONAL THEORY WORTH CONSIDERING

These are the five major areas of evidence persuading the two Soviet scientists that our Moon might be a spaceship.

Although this theory is unquestionably intriguing and even perhaps to some extent plausible in some respects, the evidence the Soviet scientists marshaled in support of it is hardly compelling.

As the two Soviet scientists concluded their own scientific paper in *Sputnik:*

> We have put forward in this article only a few of the reasons—unfortunately the evidence is so far only circumstantial—for our hypothesis, which at first glance may appear to be crazy.
>
> A similar "crazy" idea was put forward in 1959 by Professor Iosif Shklovsky, an eminent scientist [also connected with the Soviet Academy of Sciences], in relation to the "Moons" circling around Mars. After carefully weighing the evidence he concludes that they

are both hollow and are therefore artificial satellites. [Shklovsky later discarded this theory.]

We feel that the questions we have raised in connection with our Moon provide sufficient food for serious thought on the matter; the result may be the illumination of many lunar riddles.

Now, of course, we have to wait for direct evidence to support our idea. Or to refute it.

Probably there will not be long to wait.

Five years ago, when a student of mine brought to my attention the *Sputnik* article detailing this "crazy" theory, I rejected it offhand as a piece of sheer science fiction. Finally I decided the idea would be an intriguing vehicle for a science-fiction novel and began to research everything we had learned about the Moon from our space program. I soon was shocked to find out that *all the evidence both from American and Soviet lunar programs and explorations supported the crazy spaceship theory*.

So what I started to write as science fiction I ended up writing as scientific fact. Undoubtedly some people, even after reading my first book, *Our Mysterious Spaceship Moon*, still consider it to be just that—science fiction.

But after years of intensive research I still continue to discover facts and evidence that indicate clearly that the Soviet scientists are right. After reading this sequel on this all-important subject we are convinced you will agree.

Let us now proceed to devote a chapter to each of the five areas of evidence the Soviet theory delineates. Let us objectively look at the latest findings garnered from both lunar programs and let the scientific evidence fall where it may. I think, however, that you will agree with me that, surprising though it may be, the evidence in favor of the spaceship theory is impressive—enough to warrant serious consideration by the scientific community of what has to be the most far-out, most science-fictionish theory ever conceived by the mind of man.

But deep students of life recognize that invariably the science-fiction theories of yesteryear have become the scientific facts of today.

This someday may be the case with this mind-boggling and far-reaching theory of Spaceship Moon.

One asked why men should think there was a world in the moon? It was answered because they were lunatics.

— *Jest Book*

Now, my suspicion is that the universe is not only queerer than we suppose, but queerer than we can suppose.

—*John B. S. Haldane*

FIVE
MAN'S OWN SPACESHIP WORLD?

Undoubtedly most people when they first hear about this theory will think that its authors must have rocks in their heads. But such a startling theory should not be automatically rejected just because on the face of it it seems impossible.

Let us illustrate. Consider this seemingly wild theory that many NASA scientists strongly believed in before man went to the Moon. Would you believe that any scientist connected with our Moon program actually believed that rocks from our neighboring world existed on Earth *before* our astronauts brought some back here to our planet?

Dr. Dean Chapman did, and so did many of his outstanding colleagues at NASA. They were convinced "Moon rocks" had come to Earth long before the Apollo astronauts brought any here. Crazy, you say?

Chapman could demonstrate quite convincingly that certain small blobs of black glassy rocks called tektites, which are found scattered from Australia to Southeast Asia, actually splashed to Earth when the great Tycho Crater was formed by a gigantic asteroid in recent selenological times. Chapman is no amateur but a professional scientist—an aeronautical engineer who is an expert in the field of space. He helped design the heat shield that protected the American astronauts' reentry into Earth's atmosphere on space flights.

Many other outstanding scientists, including Dr. Gene Shoemaker of NASA, agreed with Chapman's basic theory.

"The earth is strewn with lunar debris," claimed Shoemaker. A meteorite striking the Moon at tremendous speed can hit with such force that it throws Moon rocks and dust into outer space, away from the light, one-sixth-gravity pull of the Moon. This Moon debris drifted freely through space, until our larger Earth swept up these "sprays of moon stuff," as Shoemaker calls it, and they eventually landed here to become part of the planet.

Before our manned lunar program got underway our space agency actually made great efforts to find such "Moon meteorites." Dr. Hugh E. Horton of the Manned Spacecraft Center in Houston tells a story about this program, which was called Project Moon Harvest. One of the places they concentrated their search on was the normally rock-free agricultural fields of the Midwest.

One investigator was looking for rocks in these fields one day when he was asked by a curious old farmer what he was looking for. The scientific researcher answered: "We're looking for Moon rocks—that's what you might call it."

The farmer flashed a smile of ridicule and replied: "Well, fella, have you looked inside your head lately?"

Undoubtedly many people will say the same about this Soviet spaceship theory—they've got rocks in their head, Moon rocks! But as we shall see, anyone who closes his mind without at least considering the evidence—and there is much—will miss perhaps the most significant scientific discovery of modern times, if not indeed of all time!

Admittedly, the author himself was guilty of unwarranted skepticism. I was convinced that the manned Apollo trips, which our scientists had hoped would solve once and for all the riddle of this rock in our skies, would furnish us with all the answers. Surprisingly, they turned out to spawn only more mysteries. And, shockingly, all these mysteries uncovered by NASA scientists and indeed scientists the world over are understandable only in terms of the Soviet spaceship theory.

Still, I remained skeptical. Hollowing out and converting such a huge planetoid as our Moon into a spaceship that could traverse the universe seemed to me beyond the pale of believability.

THE WORLD OF THE FUTURE?

While it is true that the idea of a huge, hollowed-out spaceship fashioned by human ingenuity and technology into a vast cosmic ark and sent to travel through the universe is not new among science-fiction writers, the idea that any group of intelligent beings could actually restructure a world the size of the Moon was totally beyond my comprehension.

Surprisingly, however, I have learned that our scientists have been writing for an amazingly long time about this very possibility: someday journeying to the stars in such "inside-out" worlds.

Among the first in our century to come up with such a mind-boggling idea was Professor J. D. Bernal, who back in the twenties, in his classic work *The World, the Flesh and the Devil,* suggested that entire colonies of human beings might one day be hurtling through the universe in such giant arks, living in completely enclosed inside-out-type worlds on an endless odyssey through space—just as we travel on our Spaceship Earth, only not locked to one particular star, in endless circuit around our Sun.

Some imaginative science and science-fiction writers, such as Arthur Clarke, are convinced that man's destiny lies out there among the stars, actually believe that someday such worlds may become a commonplace reality.

In fact, a few have even been bold enough to suggest that in the distant future more humans will spend their existence journeying through the universe in such spaceship worlds than have so far sojourned on Spaceship Earth. Hundreds of billions, in their view, will be born, live, love, and die in such worlds—more than the total previous human population that has existed on Earth.

Back in the fifties famed science-fiction entrepreneur Hugo Gernsback's intriguing magazine *Science Fiction Plus: Preview of the Future* carried an article (April 1953 issue) written by a Dr. Leslie Shepherd, scientific advisor to the British Interplanetary Society, who projected the possibility of fashioning such a thousand-year ark from a huge asteroid. Inside such a spaceship world entire generations of humans would voyage among the stars, "a voyage whose end would be seen only by generations yet unborn."

Isaac Asimov, well-known science and science-fiction writer, put forward similar ideas in a magazine interview about man's future. (*Strange Stories, Amazing Facts.*) Asimov claimed that it is almost certain that man will spread his kind through the galaxy—if we survive the critical problems that face our Space Age world. Someday, asserts Asimov, we will have to go—in the distant future our Sun will become unstable and flare up into a nova.

"The most practical way, barring new kinds of technical advances that are hard to foresee, is to build large ships that are self-contained ecologies—small, self-contained worlds—and just send them off, not with any special destination in mind. Every once in a while, one of them will come across a world that can be colonized."

Asimov even thinks that such worlds will (as we have noted) be made out of asteroids. This space optimist believes that man will eventually colonize millions of worlds as he travels throughout the universe in such hollowed-out spaceships. He states unequivocally that man will encounter other intelligent beings in these endless space odysseys.

A SPACESHIP WORLD CIRCLING JUPITER?

Shockingly, Arthur Clarke believes that he has discovered such a spaceship world in our own solar system. In his book *Voices from the Sky* (Harper & Row, 1965) he reveals:

"Dr. Shklovskii's stimulating theory [that a moon of Mars is hollow] appeals to me because some ten years ago I made an identical suggestion concerning the innermost moon of Jupiter. In a story called 'Jupiter V' I pointed out certain peculiarities of this satellite and developed the idea that it was a giant spacecraft, which ages ago had entered the solar system and then been 'parked' in orbit around Jupiter while its occupants went off in more conveniently sized vehicles to colonize the planets."

Apparently the Soviet scientists Vasin and Shcherbakov are not the only researchers who have suggested such a bold theory. Yet despite these claims, personally we found their assertion that an alien technology could have hollowed out such a huge orb as our Moon was just too incredible to believe. So we remained skeptical.

Then I discovered that our own scientists have conceived of building a similar spacecraft for travel to distant stars someday. And our space engineers have even formulated plans and written books promulgating this idea of futuristic interstellar space travel. One of America's most imaginative scientific minds, Dandridge Cole, in his mind-boggling book *Islands in Space* (Chilton Books, 1964; written with Donald Cox), suggests that the vast distances and time involved in projected star travel could never be accomplished through the cumbersome, limited "horse-and-buggy" type of vehicle, the rocket ship.

However, there is a way, Cole suggests, for man someday to get to the stars. His plan is to capture a huge asteroid or planetoid ranging near Earth and, with the technology that would be available to us in the near future, fashion it into an artificial, hollowed-out world.

So it is not beyond the pale of possibility, I discovered, that some superior and advanced race of alien beings capable of a greatly advanced technology—obviously vastly superior to any we have at our present stage of evolution—would have hollowed out this huge orb and turned it into a similar unique inside-out world, then powered their new cosmic home, filled with farms and factories, schools and homes, and whatever else would be necessary for their own peculiar civilization to begin the endless trek through the universe. Undoubtedly they fortified the unique world spaceship which was to be their home for generation upon generation, for—who knows—maybe century upon century, even millennium upon millennium.

Fortified it? Certainly, for the gigantic meteors, asteroids, and other celestial missiles which they would undoubtedly from time to time encounter as they traveled through space could have wrecked and destroyed their unique spaceship world. So in their fortified, metallic inner hull, with their spaceship filled with the necessary provisions in an enclosed, self-cycling, self-sustaining world, they headed their artificially constructed sphere—which to all outward eyes would appear to be just a huge asteroid—through the trackless realms of space, on what must certainly have been one of the greatest odysseys ever conceived or taken in the entire universe!

If the Soviet spaceship theory is correct (and we shall soon see that all the Apollo evidence backs it up), an unknown intelligent race of beings did precisely this billions of years ago. It staggers the mind to believe there existed beings capable of such a greatly advanced technology—especially that long ago. Obviously, even then, at the outset of their journey, they were vastly superior to us today. It could be they were forced to create a home for themselves when their sun began to die (as all stars must) or their planet home was faced for some unknown reason with extinction. Such an eventuality would lead us to take a similar step.

How did they accomplish the feat of making a spaceship world? The same way, probably, that Cole suggests for our own space program. They undoubtedly captured this huge planetoid (or perhaps one circled their planet already). Then they proceeded, with a technology that we as yet cannot divine, to hollow out this huge orb and transform it internally into a unique inside-out world, perhaps something like what Cole envisions. (See picture.)

Then they took their new cosmic home on their great star trek through the universe. Most importantly, the Soviet scientists are convinced they fortified their inner home with some tough inner hull—probably made out of alloys of metal of some kind. Naturally, to withstand the rigors of their great space odyssey that spaceship hull would have had to be super-strong. And what are the super-tough elements known to man? Titanium, chromium, zirconium—to name three. They are corrosion- and heat-resistant elements. Though they are comparatively rare on Earth, they are surprisingly plentiful on the Moon. Why? It could be as the Soviet scientists theorize: evidence of their artificially reinforced hull and surface shell.

It is interesting that we are just now using more and more of these materials ourselves because of these admirable qualities. Not only for lining electrical furnaces, but for supersonic jets and spacecraft. Zirconium, also plentiful on the Moon, though a rare element on Earth, is strikingly unaffected by neutrons, so plentiful in atomic wastes.

Is this proof that indeed a super-strong hull exists inside the Moon, made out of these super-strong elements, which lunar alien beings used to fashion and strengthen their spaceship world? Hardly. But as we shall see, rather convincing evidence indicates that such a hull exists inside the Moon.

The major point here is that the very fact that some of our own scientists and engineers are conceiving and planning to build a similar spaceship world to travel through space to the stars makes the Soviet spaceship theory suddenly plausible.

A common reaction to our first book delineating the Soviet spaceship theory was the objection: "Why would any beings go to such lengths to build such a huge spacecraft? It seems preposterous."

To answer this objection we must point out that for interstellar space travel one must face the fact that the dimensions of the universe are incredible. Light, the fastest entity known to man, can traverse trillions of miles each year. Yet, for all its speed, it takes light 4.3 years to trek the vast gulf that separates us from the star nearest Earth, Alpha Centauri—a distance of 25,000,000,000,000 miles!

Most scientists at NASA are convinced that this probably is farther than man will be able to travel in the foreseeable future. And, some are convinced, farther than man ever will.

Many scientists doubt that man will ever conquer interstellar space to reach even the closest stars. The distances are just too vast. How would one ever develop a rocket ship that could hold enough food, fuel, and other supplies for such an immense journey? It would have to be a cosmic Noah's Ark whose dimensions would be on the order of our own Earth or Moon.

Precisely. Our Earth is a spaceship in a sense, moving through a tiny corner of the cosmos. Only our planet is locked in orbit around the Sun, an average star which is also traveling through the universe.

The planet Earth is like a dog on a leash. Perhaps the planetoid Moon was once a similar world.

Another important problem frequently brought up in interstellar travel considerations is the psychological stress of any long voyage through space. But to travel through space as we do now, on a world similar to our own, or perhaps inside a world, would eliminate this problem. For the two spaceships would be very similar—except on the inside-out world the horizon would rise above one instead of curving away, as it does on our exteriorly inhabited Spaceship Earth.

The more one thinks about the plausibility of Spaceship Moon, the more one becomes convinced that what appears to be science-fiction instead turns out upon examination to be based more on scientific fact.

If you find yourself flinching at this wild suggestion, you should call to mind the words of the great Niels Bohr, the father of atomic energy, who once observed, when told by a colleague about a way-out, lunatic theory:

"Your theory is crazy. But not crazy enough to be true!"

Perhaps this Soviet spaceship theory is!

When I say a thing is true, I mean that I cannot help believing it. I am stating an experience as to which there is no choice.

—*Oliver Wendell Holmes*

SIX

PREDICTIONS BASED ON THE SPACESHIP THEORY

The two bold scientists who postulated the spaceship theory ended their thesis with this statement:

> We feel that the questions we have raised in connection with our Moon provide sufficient food for serious thought on the matter; the results may be the illumination of many lunar riddles.
>
> Now, of course, we have to wait for direct evidence to support our idea. Or to refute it.
>
> Probably there will not be long to wait.

In this fashion Alexander Shcherbakov and Mikhail Vasin, Soviet researchers, concluded their case.

In our upcoming chapters we shall begin to take a look at their theory, examining the evidence that has poured back from the lunar programs, both Soviet and American. But before we delve into the evidence to see if it backs up the spaceship theory or wrecks it, let us do a little "lunar" armchair quarterbacking. If you had read this shocking treatise in the Soviet publication *Sputnik* back in July of 1970, as I did—in which two orthodox scientists claimed that our Moon was not a completely natural world but a huge, hollowed-out planetoid; that it had an inner shell of metal which served as its hull; that it had a reconstructed outer shell which was partly metallic rock; that there was evidence of artificial construction in its interior; that after traveling as a spaceship for eons through the cosmos it was powered and "steered" into orbit around our world at some undeterminate time in the past by some unknown aliens— your mind too would have boggled.

If at that time you were a lunar expert with an open mind anxiously awaiting proof or disproof of this wild theory, you could have sat down and predicted what scientists should find on the Moon before our astronauts completed their lunar voyages. You can do it now.

As you prepare to make your own Moon odyssey through this book, check these commonsense "predictions" of what scientists could expect to discover about our Moon. See if they do not come true. You will be amazed at how Spaceship Moon lives up to expectations, how it appears to show the very characteristics that one would expect from a hybrid world—a natural asteroid "constructed" and converted into a spaceship!

Since our two Soviet spaceship theorists claimed that the Moon is a huge, hollowed-out world, then the following should have proven true:

(1) If it really is hollowed out and a natural asteroid restructured into an inside-out world of a spacecraft, then orthodox scientists could be totally confused by what they find. Nevertheless, despite discovery of seemingly contradictory facts and evidences, it should definitely show clear proof that it is a hollow world. However, because

of all kinds of artificial construction in the interior, the data supporting the hollow-Moon contention might seem somewhat confusing and to some extent be masked. Density studies and gravitational field studies nevertheless could give us tantalizing clues of what most orthodox scientists consider to be impossible—a natural hollow satellite.

(2) If this alien asteroid was refashioned into an artificially hollowed-out spaceship, then it should show evidences of massive, intense melting, undoubtedly many miles in depth.

(3) A corollary of this: Some scientists, such as Dr. Harold Urey, were convinced before man went to the Moon that it must have been formed "cold"—that it was too tiny an orb to have generated sufficient heat to have been a hot body in space and produced extensive melting or volcanic eruptions to the exterior. Other scientists disagreed, claiming that evidence pointed definitely to a "hot" Moon. All scientists felt that going to the Moon would prove conclusively one way or the other which theory was correct.

However, if the Moon was a natural asteroid which had been artificially restructured, then both sides should find evidence to support their theories. Both should be able to garner facts and evidence each contradicting the other, yet each supporting their respective orthodox positions. Unless, of course, both consider the possibility that they might be looking at an artificially restructured spaceship. Then seeming contradictions become suddenly enlightening evidence.

(4) If these aliens did pour out much internal lava, either to hollow out or later to reinforce their spaceship, then the outside surface areas of the Moon would contain material that originally was deep within the body of this world. This should give a strange picture to lunar experts, making it appear as if the Moon had been made "inside out." Moreover, if the alien used metallic elements in reconstructing or reinforcing the outer shell, the lunar samples that the astronauts brought back might contain pure metal or metal particles with evidence of artificial processing or manufacturing, perhaps even giving evidence of a superior technology beyond even that of modern Space Age man.

(5) If the moon is a huge hollow sphere and if the interior of this world really has a metalic-type spaceship hull, as the Soviet scientists claim, then seismic results of Apollo

spacecraft crashed into it (such as lunar modules and rocket stages) should produce different, even puzzling vibrations and tremors. If the hull is metallic these tremors could be of extremely long duration. In fact, these man-made crashes could make this huge metallic sphere vibrate like a huge bell or gong. Also, the vibrations should carry extremely far, being conveyed not just great distances but, if the hull goes all the way around the Moon, as the Soviet reseachers say, carried completely around the entire orb. Unheard of in a natural world, but something to watch for.

(6) If these alien "Moonmen" used materials such as iron and titanium in creating this inner hull or shell, as well as in forming the maria, then large areas of the Moon should be devoid of metals while inner areas and the metal-rich maria should prove to have them in great amounts.

(7) If high-temperature metals were used in these constructions, which require intense heat to melt, the Moon might give evidence of this. What could alien beings have used to produce such high temperatures? Probably extensive concentrations of radioactivity, properly used. If this is true, then the Moon may give evidence of being much more radioactive than Earth. Secondly, since this lava was poured out onto the surface of the Moon, these outer areas should contain high concentrations of radioactive elements, maybe even intense hot spots. This would appear to be puzzling to orthodox scientists, but evidence of this alien formed world that our Soviet theorists posit should be found on the moon.

(8) If these Soviet scientists are correct in their surmise that the Moon as an artificially formed spaceship came here from somewhere else, then the Moon will present a different face from what scientists expect. Its bulge, for instance, may even be on the side that has never faced us—the back side of the Moon—instead, as scientists have always assumed, on the near side!

(9) If the Moon has an internal atmosphere in its hollow regions, probably to sustain life, then the outer surface might be extremely dry but at times could give evidence of a great amount of water or other gases. The Soviet scientists theorize the inner part of this alien world is probably filled with "gases required for breathing, and for technological

and other purposes," and some of this gas might be vented outward "through cracks appearing in the armour plating." If so, these gases could be detected outside, on the surface. They could very well have a high water vapor content, so that this dry, dry Moon might yet show surprising amounts of water vapor.

(10) Similarly, although earlier Moon probes indicated that the Moon has almost no magnetic field, if the Moon did travel through the cosmos as long and as widely as the Soviet spaceship theorists think it did, then Moon rocks might have been affected by other global and solar magnetic fields and evidence of this would be "frozen" into Moon rocks. Our scientists could therefore be in for a big magnetic surprise!

(11) If the Moon did come from afar and has not always been circling Earth, then the make-up of the Moon, its composition, could be much different. It might even contain elements our planet Earth does not have! Moreover, it might give evidence of being older than the oldest rocks we have found on Earth. Finally, if it traveled through different parts of the universe, trekking through different star systems of differing ages, then its surface should be covered with different rocks and particles of widely varying ages. And it is even possible that if those portions of the universe were younger than the original "home" of the aliens, we could have lunar crust being much older than many of the rocks lying on its surface! What to scientists would appear to be an impossible order of things.

(12) Finally, if the Moon is such a refashioned planetoid or asteroid with all kinds of artificial construction in its interior, then seismic records would probably yield some perplexing results—not the least perplexing of which might be identical signals trackings. For artificial construction might on the Moon's interior give a seismic picture that could be virtually the same each and every time. On the surface, to scientists not tuned into the artificial-Moon hypothesis, this might appear impossible.

In conclusion, lunar experts might find many puzzling, contradictory conundrums on such a world. Many impossible findings and discoveries could lead scientists in general to find the Moon more of a mystery now than before we examined it at close range. Because of such contradictions and conflicts, scientists might find it impossible to weave a unified theory that would explain this mystery world, the Moon.

But in light of the spaceship theory the mysteries would dissolve. But among orthodox scientists, overwhelmed by the multiplying mysteries and perplexing problems, a cynical few might conclude that such a "crazy" world cannot exist. For without the spaceship moon hypothesis, they will not be able to explain its origin or its nature; they will not be able to begin to understand it.

By now, you know how strange a place the Moon is, how it should not even be here. Its origin is even stranger.

—George Leonard

SEVEN
WAS OUR MOON "POWERED" AND "STEERED" INTO ORBIT AROUND OUR EARTH?

Consider these tantalizing questions:

• Why is it that how the Moon came to our skies remains "the biggest puzzle of them all," according to NASA's own testimony?

• Why do leading scientists confess that they do not know where the Moon came from or how it got here—that, as one Nobel prize-winning scientist admitted, "All explanations are improbable"?

• Why does one NASA expert admit: "It seems much easier to explain the non-existence of the Moon than its existence"?

• Why do our leading scientists today facetiously suggest that the Moon *does not exist at all,* since there seems to be no way for it to have originated?

- Why do some leading scientists refer to the Moon as a *cosmic freak of nature*—claiming that it is too big, too far out to be the natural satellite of Earth?

- What evidence exists that our Moon is not the true, natural satellite of Earth?

- Why are scientists today unable to comprehend or explain the most commonly accepted theory among lunar experts, which claims that our Moon was naturally "captured"?

- Why do most experts in celestial mechanics claim that the capture of our Moon in the gravitational field of our own planet is virtually impossible or at the very least highly improbable?

- How does our Moon's stable, well-behaved circular orbit indicate that it was not naturally put into orbit around our world?

- Why do some leading scientists claim that the Moon was put into such an orbit by "some unknown force"? What is that force?

- Why is the Moon so precisely positioned in the heavens that to an observer on Earth its disk size exactly equals the Sun's disk size? How does this make eclipses possible—the only planet that we know of that has them?

- Why do scientists find it incredible that our Moon has such a precise position and why is this not understandable except through the Soviet Spaceship Moon theory?

- Why does a knowledge of celestial mechanics and the experience from our own space launching of satellites indicate that the Moon was "powered" and "steered" into orbit around the planet Earth?

Now let us turn to the evidence...

THE MYSTERY OF THE MOON'S ORIGIN

Victor Hugo, the great French author, once called the Moon "the kingdom of dreams, the province of illusion."

It has been just that for scientists. They figured that the Moon could very well be the Rosetta Stone of our universe, the key that would unlock the secrets of the origin of our solar system as well as furnish many, many secrets of the cosmos. That was before we went there.

So far it has been anything but that. So far the Moon has only served to confound and perplex scientists. For we are now more confused about its origin and its nature—about where and when and how the Moon came into being and what its make-up essentially is—than we were before we went there.

THE MOON—A PRE-APOLLO PUZZLER

Admittedly, before man traveled to our nearest neighbor in space, all science was in a quandary about this strange, enigmatic world circling our Earth. Where it came from and how it got here were questions our scientists had no certain answers to.

Three major theories were developed by scientists of Earth to explain how the Moon came into being—or at least how it came to be circling Earth. First the hypothesis was advanced that it evolved along with Earth, both being born out of the same cosmic cloud of dust and gas about 4.6 billion years ago. The second theory is that our Moon fissioned or split off from Earth after our planet's birth eons ago. The third is that the Moon was a latecomer, formed independently and far from our Earth, then "captured" and locked into orbit around us as it was passing through our planet's gravitational field.

The majority of Earth scientists thought that our Moon was formed with the planet out of a cosmic cloud of dust and gas about 4.6 billion years ago. A smaller segment of the scientific community thought it might have once been a part of our planet. George Darwin, son of the famed evolutionist Charles Darwin, theorized that our Moon was split off from our planet in an act of celestial fission. Famed astronomer William H. Pickering adopted a form of this theory when he theorized that it might have been torn out of the Pacific area early in our Earth's evolutionary career.

Although theoretically this vast Pacific basin is large

81

enough to hold a body as great as the Moon, and although there does exist around the rim of the Pacific a volcanic, earthquake "ring of fire," believed by some scientists to be evidence of the birth pangs of Luna leaving the womb of Mother Earth, most researchers rejected this theory, since immense difficulties of celestial mechanics made it nearly impossible. The young age of the Pacific itself militated against this kind of fissioning off of our Moon, for this would have had to have taken place at an early stage of Earth's evolution. On this all scientists agree. Thus, it could not have been wrenched from this area even if scientists were to assume the celestial mechanics were possible.

Furthermore, the death blow appears to come from the Moon itself. Evidence extricated from our Apollo expeditions indicates that the composition of the Moon is much different than that of Earth. If the Moon were a child of Earth, fissioned off from the Pacific (or any other area), then we would expect its composition to be about the same. It is not.

These very same findings seem to militate against the Moon's evolving with our own planet and forming out of the same cosmic cloud of dust and gas. And the fact that further evidence indicates that the Moon may be older than Earth also deals this theory a crushing blow.

This leaves only the capture theory, an idea that was not too well accepted by scientists before Apollo, since astrophysicists and experts in celestial mechanics claimed that such an eventuality was highly improbable, if not virtually impossible.

Even if scientists were to assume that the improbable capture could have happened, that the Moon would have ended up with its present circular orbit is just too incredible to believe. For a natural celestial capture—and this astrophysicists stress—would have produced a highly elliptical instead of a circular orbit. How then to account for the Moon's circular orbit around our Earth? No scientific theory adequately does—except, of course, the unorthodox theory of two orthodox scientists of the Soviet Academy of Sciences who maintain that the Moon was powered and steered into orbit around Earth.

WHERE THEN DID THE MOON COME FROM?

"This is the biggest puzzler of them all," observes a NASA document. And one to which not only NASA scientists but experts the world over freely admit there is at present no solution—despite all our efforts, all our data, facts, and figures.

The truth is that today scientists are unable to come up with any adequate theory that embraces all the known, and often apparently contradictory, facts about our Moon. No one has discovered a way to solve all the conundrums and perplexing puzzles. No one has worked out a theory that will unify all the conflicting data, molding them into a sensible hypothesis. No one save the Soviet spaceship theorists, that is.

Everything points, however, to the fact that our Moon came from elsewhere in the universe, as we shall see. One of the few scientists to hold that this was so even before Apollo was Dr. Harold Urey, Nobel Prize winner.[*] Urey held that precisely because our Moon was from another part of the cosmos, the Moon would necessarily be as fascinating an alien world as any that man could find. As Urey put it: "Stepping on the Moon would have the same interest as stepping on Mars, or the asteroids, or Venus."

We point out that if the Soviet spaceship theory is correct then it would be of even greater interest! The greatest one that man could possibly imagine!

DOES THE MOON EXIST AT ALL?

Unbelievably, this has come to be a joke among lunar scientists today. Why go that far, even if it is in jest? Because, as Dr. Robin Brett notes, NASA scientists "weighing

---

[*] Back in the 1960s Urey asserted: "Is it possible that two bodies like the Earth and the Moon could have accumulated near each other in space from debris of some kind, and have markedly different densities. So far as we know there is no reason why the more dense material should accumulate into a large object, the Earth, and the less dense material should prefer to accumulate in the small object. Thus, accordingly as an explanation for two planets so near each other appears to us today to be very improbable."

what has been learned" do not seem to be able to account for the existence of our Moon.

"All three theories have weaknesses in the light of our present knowledge," he explains. "The composition of the returned lunar samples makes it difficult to derive them from anything like the composition of the Earth's mantle. This, therefore, makes the fission theory extremely unlikely. And if the Moon was formed as an identical twin planet with the same composition as the Earth's mantle, the same argument applies against that theory. The capture theory presents difficulties in celestial mechanics and is regarded as statistically improbable. *It seems much easier to explain the non-existence of the Moon than its existence.*" (John Wilford, *We Reach for the Moon*, W. W. Norton, 1971. Emphasis added.)

Dr. William K. Hartmann, senior scientist at the Planetary Science Institute in Arizona, points out in *Astronomy* magazine: "Each of these main pre-Apollo themes seemed to have a fatal flaw. Some scientists were driven by frustration to facetiously suggest that *perhaps the Moon doesn't exist at all, since there seemed no way for originating it.*" (Emphasis added.)

IS THE MOON A COSMIC FREAK OF NATURE?

Many scientists in modern times have come to regard the Moon as a cosmic freak of nature. Why? Because by all cosmic laws, they point out, *the Moon should not be orbiting our planet Earth!*

Why do leading scientists like Isaac Asimov assert this? Simply because our satellite seems to be too big for our planet earth. It is a whopping one-fourth of our planet's size, and proportionately is the largest satellite that we know of orbiting any world. For though a satellite of Jupiter, for instance, is actually larger in size, the Jovian satellite Ganymede is only one twenty-seventh the size of the planet it circles.

THE MOON IS TOO FAR OUT

Not only is our Moon too large, but scientists also point out it is actually too far out in its orbit to be a natural satellite of our planet. The prolific science writer Isaac Asimov, who is a scientist in his own right, asks: *"What in blazes is our own Moon doing way out there?"* (Emphasis added.)

"It's too far out to be a true satellite of the Earth. It's too big to have been captured by the Earth. The chances of such a capture having been effected and the Moon then having taken up nearly circular orbit around our Earth are too small to make such an eventuality credible." (*Asimov on Astronomy*, Doubleday, 1974.)

Scientists in general would agree. But most remain confounded by this conundrum. Our Moon should not be there in circular orbit around our Earth, and many insist it should not be there period! Not perhaps as a natural world, but an artificially "steered" world, yes. Grudgingly all scientists must admit that the problem disappears in light of the Soviet spaceship theory.

THE MOON NOT A TRUE SATELLITE OF EARTH?

Isaac Asimov concludes that our Moon is not, as is commonly believed, the true satellite of our planet. He claims that if it were "it would almost certainly be orbiting in the plane of Earth's equator and it isn't."

There are still other powerful reasons why the Moon cannot be a true satellite of Earth, insists Asimov. He claims the Moon should not be orbiting our planet because of what he calls "the tug of war" ratio.

"The Sun attracts the Moon twice as strongly as the Earth does." (*Asimov on Astronomy*.) This ratio of the Moon's size and distance in relation to the Earth and to the Sun indicates that it should have been gone long ago. The Sun should have won this tug of war and pulled the Moon away, but somehow it hasn't.

Asimov claims that our so-called natural Moon gives every indication that it is not a true satellite of ours. He also rejects the possibility of its having been captured. This seems to lead to the horns of a dilemma, as he himself points out:

"But, then if the Moon is neither a true satellite of the Earth, nor a captured one, what is it?" (*Asimov on Astronomy*.)

Asimov tries to slip off the horns of this dilemma by postulating that the Moon is really a planet in its own right. The problem here is that all the evidence from our Apollo and the Soviet flights indicates that our Moon is too different in make-up to be a double planet, so this does not appear to be the correct theory either. Further, if evidence that the Moon is older than the Earth is correct, then it must have come here from somewhere else in the universe. That brings us back to the only other possibility (aside from the spaceship theory): namely, that the Moon was captured. And this leads us back to the problems of how it came to be captured by our Earth and how it could have ended up in its present circular orbit.

Surprisingly, more and more scientists, despite all these difficulties, are now considering this theory that the Moon and our Earth somehow evolved separately, far from each other, and our Moon somehow came to be captured by Earth. But this is hard to accept, not only because of the previously raised objections but also because of the unique, strange position the scientists find the Moon in today. And here we may discover another hint that indicates that our Moon did not "just happen" to "fall" into a chance orbit around our planet.

A REMARKABLE COINCIDENCE OR EVIDENCE OF ALIEN PRECISION?

The disks of the Sun and the Moon appear to be just about equal as viewed from Earth. Of course, the apparent size of the Moon and the Sun depend upon their respective distances as well as their actual sizes. The Moon is only 2160 miles in diameter, while the diameter of the Sun is 864,000 miles. That makes the Sun's diameter approximately 400 times greater. To put it another way, the ratio of our Moon's disk to the Sun's is 1:400.

However, our Sun is 93,000,000 miles away and the Moon only about a quarter of a million miles away. Strangely enough, this works out to about the same ratio—approximately 1:400. So the distance just about cancels out the size, and this is why the tiny Moon appears to the viewer on the Earth to be about the same size as our gigantic Sun.

This is shown remarkably during a total eclipse. Isaac Asimov makes an observation about this truly amazing situation, which he calls "coincidence."

In his book *Space, Time and Other Things* (Doubleday, 1965) he notes: "What makes a total eclipse so remarkable is the sheer astronomical accident that the Moon fits so snugly over the Sun. The Moon is just large enough to cover the Sun completely (at times) so that a temporary night falls and the stars spring out. And it is just small enough so that during the Sun's observation, the corona, especially the brighter parts near the body of the Sun, is completely visible."

What does Asimov make of this remarkable piece of astronomical good luck? "There is no astronomical reason why the Moon and the Sun should fit so well. It is the sheerest of coincidences, and only the Earth among all the planets is blessed in this fashion."

THE SHEEREST OF COINCIDENCES?

The chances of the Moon's being in a position so exact as to just equal the disk of the Sun in relation to the planet Earth is—pardon the pun—astronomical. To paraphrase a quote from Asimov, which he uttered about the possibility of the Moon's having been captured by Earth and ended up in a circular orbit: *The chances of the Moon's just happening to be in this position are too incredible to be credible.*

This "accident" is just too astronomically farfetched to be truly an accident. In light of the Spaceship Moon theory, however, all these striking aspects of our Moon's strange orbit around us—its circular, nonequatorial orbit plus the other strange characteristics—seem to make it clear that our nearest neighbor in space is where it is not by accident.

Admittedly, if the Spaceship Moon theory is accepted, all these problems disappear. If intelligent beings put Spaceship Moon in orbit around Earth—which we are confident will soon become clear and convincing to you—then the reason why the Moon lies in precisely this "remarkable" orbit is clear as day. To sum it up simply: It was "put" there!

One physicist at a Midwestern university, who has become convinced that our Moon may well indeed be a spaceship, remarked on this strange position of the Moon: "It's almost as if these aliens were waving a flag at man and saying—*'Look here! Here is proof that the Moon is more than just a mere moon!'* "

NO OTHER MOON MODEL IS CORRECT

But despite this intriguing evidence, the vast majority of orthodox scientists adhere to the old theories. Without the spaceship theory to resolve the kind of difficulties we have pointed out, the majority of scientists at the last major world meeting of lunar experts, The Fifth Lunar Conference, officially adopted the capture theory that holds that our Moon was gravitationally "kidnapped" by Earth. They didn't resolve the difficulties of the celestial mechanics; they just went ahead and adopted it. Before man went to the Moon the vast majority of these scientists would not begin to even consider this far-out hypothesis because of the immense difficulties it entails.

Dr. Harold Urey summed it all up when he wrote before the Apollo journeys: "All explanations now offered are improbable." He holds that to this day. "I do not know . . . the origin of the Moon, I'm not sure of my own or any other's model. I'd lay odds against any of the models proposed being correct."

Urey himself had leaned toward the capture theory, although he too had serious reservations about it. For even if scientists were to grant that the Moon, as a wandering orb of the outer cosmos, just happened to drift too close to our planet's gravitational field, and even if all conditions were proper for it to be gravitationally caught, it is hard to see how it could have ended up in its present orbit.

Such a capture should have produced a rather elongated, elliptical orbit instead of the nearly circular path the Moon now follows around Earth. Indeed, as the Soviet scientists insist, it seems that it was steered into orbit around our planet.

A well-known American science writer, William Roy Shelton, makes this striking observation in his book *Winning the Moon* (Little, Brown & Company, 1970), which focuses in on this problem:

> It is important to remember that something had to put the moon at or near its present circular pattern around the Earth. Just as an Apollo spacecraft circling the earth every ninety minutes while one hundred miles high has to have a velocity of roughly 18,000 miles per hour to stay in orbit, so something had to give the moon the precisely required velocity for its weight and altitude. For instance, it could not have been blown out from Earth at some random speed or direction. We found this out when we first began to try to orbit artificial satellites. We discovered that unless the intended satellite reached a certain altitude at a certain speed and on a certain course parallel to the surface of the earth, it would not have the necessary centrifugal force to maintain the delicate balance with the gravity of Earth which would permit it to remain in the desired orbit. (P. 58.)

What *force* put it into orbit around Earth? If we can prove that the Moon is hollow, that is also has a hard inner shell of metal or metallic rock, such evidence would be fairly conclusive that it is indeed a spaceship. Then the problem about its peculiar circular orbit disappears.

As Shelton further delineates: "The point—and it is one seldom noted in considering the origin of the moon—is that it is extremely unlikely that any object would just stumble into the right combination of factors required to stay in orbit. 'Something' had to put the moon at its altitude, on its course and at its speed. The question is: what was that something?" (*Winning the Moon*, pp. 58–60.)

In light of the Soviet spaceship theory "that something" that Shelton speaks of is clear. In fact, we should ask not just what "that something" was, but *who* that someone was that put the Moon around Earth—for it was alien intelligence Soviet scientists Vasin and Shcherbakov claim, that put Spaceship Moon in orbit around Earth.

The author of *Winning the Moon* concludes:

"Discovering what that something was, given the moon's known distance from earth, its known huge relative size, and its known way of rotating, has occupied astronomers, scientists, and more recently, engineers in an attempt to solve one of the most fascinating riddles known to men on earth."

We are convinced that a couple of Soviet researchers have solved this great cosmic riddle. And after reviewing the array of evidence for "Spaceship Moon" we are convinced you will agree with their mind-boggling, earth-shaking theory. Indeed, after reviewing all the arguments marshaled in this book that prove the Moon to be a vast hollow world which is in fact a spaceship, you will see how clear the evidence is that our Moon was "powered" and "steered" into orbit around our planet Earth.

The moon must be enormously cavernous with an atmosphere within, and at the centre of its caverns a sea. One knew that the moon had a lower specific gravity than the earth . . . one knew, too, that it was sister planet to the earth and that it was unaccountable that it should be different in composition. The inference that it was hollowed out was as clear as day.
—H. G. Wells (1901)

EIGHT
IS OUR MOON HOLLOW?

Tantalizing questions:

• How does the Moon's strange density indicate that it may be hollow?

• Why does a leading British lunar expert, former head of the Lunar Section of the British Astronomical Association, conclude: "Everything points to the Moon being hollow 20–30 miles underneath its crust"?

• Why is it that leading scientists agree that a natural satellite cannot be a hollow object? Why can we then conclude that if the Moon's interior is hollow, it was artificially hollowed out?

• Why does a pre-Apollo NASA study done by leading NASA scientists conclude that the Moon's motions indicate that our satellite may be a hollow orb?

• How does the gravitational-field study indicate the frightening possibility that the Moon might be hollow? Why is this conclusion "frightening"?

• Why does the density of Moon rocks brought back by our astronauts indicate that the Moon is hollow?

• Why do man-made crashes of space equipment (lunar landing modules and spent stages of rockets) into the surface of the Moon cause the Moon to "ring" like a huge bell or gong? Why does the Moon vibrate for up to four hours? Why does this indicate that the Moon is a huge hollow sphere?

• Why do studies of the Moon's rotational motion indicate that the Moon may be hollow?

• What evidence is there that our own space agency carried out secret studies to see if there were extensive hollows in the Moon's interior? Why did they do this?

• Why have translational motion studies of the Moon (motion of the center of mass) which indicate that the densest regions of the Moon are nearest its surface (indicating that the Moon is hollow) been disregarded? Why are these scientific objections overcome in the hollow-spaceship theory?

• Why does all the evidence indicate that our Moon is a huge hollowed-out orb?

Let us now examine the shocking evidence!

STRANGE LIGHTNESS OF THE MOON

"Is the Moon hollow?" This happens to be the title of a chapter in *Our Moon,* one of the most authoritative books ever written on the Moon, by H. P. Wilkins, former head of the Lunar Section of the British Astronomical Association. It was published in the fifties, more than two decades before two Soviet scientists postulated their hollow-Moon theory.

Dr. Wilkins (who died at the beginning of the Space Age), one of the world's leading lunar experts, claimed that the Moon could very well be hollow—at least to a great extent. As Wilkins pointed out in the thirteenth chapter of his fine book *Our Moon: "Everything points to a more or less hollow nature of the crust of the Moon within some 20 or 30 miles of the surface."* (Emphasis added.)

Shades of H. G. Wells! And strikingly like the Soviet spaceship theory.

What led Wilkins to his remarkable conclusion?

First of all, the Moon, even before man traveled there, was known to be only about half as dense as the material out of which the planet Earth was formed. In fact, the Moon is only about 60 percent as dense as our own world. That is, if one were to take a cubic mile of the Moon and compare its weight with that of a cubic mile of Earth, the latter would weigh nearly twice as much.

Why? The fact that a given volume of Moon material seemed to weigh about half as much as an equal volume of the Earth mystified scientists. How to account for the difference?

Some of the scientists felt that the answer lay in the fact that the material out of which our Moon world was formed was in fact very light and hence weighed only about half as much.

Some scientists, such as Dr. Harold Urey, felt that the solution might be that the Moon was without a heavy core, resulting in a lower overall density. Other equally prestigious scientists, such as British astronomer Wilkins, opted for a hollow moon—at least one with huge hollow areas.

In his authoritative study *Our Moon* the eminent British

scientist explains his astounding conclusion: "Long ago it was calculated that if the Moon had contracted on cooling at the same rate as granite, a drop of only 180F. would create hollows in the interior amounting to no less than 14 millions of cubic miles." According to Wilkins, it appears that our Moon has a great portion that might contain naturally hollow caverns or regions. He adds: "However, it is unlikely that the Moon contracted at the same rate as granite; it is almost certain that nothing like 14 millions of cubic miles of cavities were formed. . . . Nevertheless, everything points to the more or less hollow nature of the crust of the Moon . . . within some 20 or 30 miles of the surface. It thus appears that hidden from us are extensive cavities, underground tunnels and crevasses, no doubt often connected with the surface by fissures, cracks or blowholes."

Later we shall see that evidence for such openings into a hollow moon does exist.

Wilkins continues:

"Without going so far as H. G. Wells, who imagined a strange race of creatures, the Selenites, inhabiting these hollows, which were however subordinate to an elaborate system of artificial tunnellings, the cavernous interior of the world within the Moon must be a strange place. Immersed in impenetrable darkness and absolute silence, the walls doubtless studded with numerous crystals, these gloomy caverns, branching and winding, here and there connected with the surface by a half choked pit or an open crack, may contain surprises for the first space-travellers to land on the Moon." (*Our Moon,* pp. 119–20.)

Not the least surprise—if we are to believe the two Soviet spaceship theorists—could be the strange home of alien beings themselves!

For if the Soviet theory that the Moon is hollow through and through is valid, then the Moon must be a spaceship! All scientists—at least all astronomers—agree with that conclusion. For from what is known about the way that worlds evolve, scientists know it cannot be naturally hollow.

Even the conservative astronomer from Cornell University, Dr. Carl Sagan, admits this. In the early sixties Dr. Sagan coauthored a book (*Intelligent Life in the Universe,* Holden Day, 1966) with another scientist connected with the Soviet Academy of Sciences, astrophysicist Iosif Shklovsky, who theorized at the time that the moons of Mars might be hollow and would therefore have to be space stations.

Sagan agreed: "A natural satellite cannot be a hollow object." (*Intelligent Life in the Universe.*)

Thus, astronomers are in general agreement that if the Moon is hollow, then it must have been hollowed out artificially! And this brings us around to the Vasin-Shcherbakov Spaceship Moon theory.

These two scientists claim that their density study of the Moon indicates that the Moon could be hollow. In their scientific treatise they postulate: "If you are going to launch an artificial sputnik, then it is advisable to make it hollow. . . . It is more likely that what we have here is a very ancient spaceship, the interior of which was filled with fuel for the engines."

Shcherbakov and Vasin observe that the density of the Moon would indicate it could be hollow: "Why is there such a big difference between the specific gravity of the Moon (3.33 grammes per cubic centimeter) and that of the Earth (5.5 gr.)?" The solution is that the Moon is hollow!

They conclude: "Since the Moon's diameter is 2,162 miles, then looked at from our point of view it is a thin-walled sphere. . . ."

They point out the precedent for their reasoning: "A similar 'crazy' idea was put forward in 1959 by Professor Iosif Shklovsky, an eminent scientist, in relation to the 'moons' circling round Mars. After carefully weighing up the evidence he concludes that they are both hollow and therefore artificial satellites."*

AMERICAN-OBTAINED EVIDENCE BACKS UP SOVIET CLAIMS

Is there any hard, solid scientific evidence to prove that our Moon is actually hollow and a spaceship?

Surprisingly, even before the Moon manned-landing pro-

---

* Shklovsky backed down from his hollow-Martian-moons theory after later evidence indicated that they might *not* be hollow.

gram was launched, a NASA scientist had come up with such evidence!

ANALYSIS OF THE MOON'S MOTION INDICATES HOLLOW MOON

It was this discovery which helped convince me that there might be something to this crazy Soviet spaceship theory. For, frankly, although I was intrigued by the Soviet hollow-moon theory and even though I admitted it might be plausible, their evidence was not that convincing. Most of it, in fact, seemed at best to be circumstantial.

About all I was convinced of was that it might be an excellent vehicle for an intriguing science-fiction novel, which I began writing. As I was researching the subject of the Moon I stumbled across some evidence that jolted me. I was reading *America's Race for the Moon* (Random House, 1962), edited by Walter Sullivan. The book, subtitled *Story of Project Apollo,* is a collection of scientific articles on the Moon and our lunar space program.

In this fascinating work I read about a study done by Dr. Gordon MacDonald of an analysis of the Moon's motions. Walter Sullivan noted in his article, "What Will the Moon Be Like":

"Even more startling is a report by a leading scientist of the National Aeronautics and Space Administration that, according to an analysis of the moon's motions, it appears *to be hollow.*" (Emphasis added.)

Digging out Dr. MacDonald's study, I discovered his conclusion:

"If the astronomical data are reduced, it is found that the data require that the interior of the Moon be less dense than the outer parts. Indeed, it would seem that the Moon is more like a hollow than a homogenous sphere." (*Astronautics,* February 1962, p. 225.)

However, as Walter Sullivan points out, Dr. MacDonald did not accept the conclusion of his own study that the Moon is hollow. This outstanding scientist, perhaps realizing the truth of the astronomical principle that no satellite world can be *naturally* hollow, that there is no such thing as a hollow planet or planetoid in space that is naturally hollow, "does not suggest that the moon really is hollow." Says Sullivan: "Rather, he believes something is wrong, either with the data or the calculations." (*America's Race for the Moon*, p. 96.)

In *Astronautics* Dr. MacDonald spells out why he rejected the hollow-Moon conclusion: "This suggests that there are inconsistencies either in the reduction of the observations of the moon's motions or in the numerical development of the lunar theory."

Thus, rather than accept the basic conclusion of the study, which indicates clearly that the Moon behaves more like a hollow sphere than a solid body, MacDonald believed that either something was wrong with the data or there were inconsistencies in observation.

The thought has since struck me time and again: What if, privately, NASA scientists and officials actually accepted the hollow-Moon conclusion and suspected that the Moon was in fact a spaceship? And that this indeed was the strong impetus to reach the Moon and find out once and for all? Buoyed by this knowledge and the evidence that strange moving lights and objects seen on the Moon by astronomical observers indicated that the Moon was the source of the myriads of UFOs flooding Earth skies, both the Soviet and the American governments, operating in secrecy, launched crash programs to reach this spaceship in Earth skies.

Of course, this is mere speculation. But not speculation without foundation—as we have seen.

MY OWN IMPETUS

When I read about the astonishing conclusion of the MacDonald hollow-Moon report I was mentally staggered. Naturally this provocative thought began to haunt me: Could the Soviet scientists be right after all? Is the Moon really hollow and thus a spaceship?

My curiosity piqued, I set to work. I read through every NASA scientific document I could get my hands on. After poring over 15,000 pages of NASA and lunar scientific studies, I discovered more than 50 major proofs that, in my view, confirm the Soviet artificial-Moon theory, including several other independent studies that indicate the Moon is indeed hollow. These range from Dr. Harold Urey's negative-mascon concept to Dr. Sean Solomon's gravitational-field study.

SOLOMON'S GRAVITATIONAL-FIELD STUDY INDICATES MOON MAY BE HOLLOW

Dr. Sean C. Solomon of M.I.T. claims that a study of the gravitational fields of the Moon indicates that it could be hollow. Solomon concludes his study, which was published in Volume 6 of the technical periodical *The Moon, An International Journal of Lunar Studies*:

"The Lunar Orbiter experiments vastly improved our knowledge of the Moon's gravitational field . . . indicating the *frightening possibility that the Moon might be hollow.*" (Pp. 147-65. Emphasis added.)

Frightening? Yes, because if the Moon is hollow it must have been artificially hollowed out by some alien intelligence—and that would necessarily make it a spaceship!

Without question, if it could be established as fact that the Moon is really hollow, it would be the greatest discovery in the history of science and the most fantastic that man has ever made.

HOLLOW LUNAR CAVITIES INSIDE THE MOON?

One outstanding lunar scientist who holds there is evidence that huge cavities or hollows might exist inside the Moon is Dr. Harold Urey. He maintains that there are "negative mascons" inside the Moon, huge areas beneath which "there is either matter much less dense than the rest of the Moon, or simply a cavity." Technical data supporting this finding were uncovered by Dr. Sjogren, codiscoverer of the mascons.

French science writer Jean Sendy, author of *The Coming of the Gods* (Berkley, 1973), is convinced such underground cavities exist: "Yes, if my underground base on the moon exists, this is where it ought to be." (*The Coming of the Gods.*) Sendy is convinced that the Moon is the home of the ancient "gods"—as well as modern-day alien astronauts!

Sendy explains: ". . . if Dr. Sjogren's study of the data transmitted by Lunar Orbiters leads to the discovery of a cavity under the surface of the Moon, and if 'my' base is found there, I will be able to say, 'You see, I was right!' And I wasn't right by accident." (*The Coming of the Gods*.)

HOW HIGH ARE THE HOLLOWS OF THE MOON?

Whether the Moon is a completely hollow, thin-shelled or thin-walled sphere or merely has huge hollow areas like cavities or caverns has not been definitely determined yet—to our public knowledge, that is. But the Moon surely behaves internally as if it were hollow, as seismic evidence clearly shows.

"MOON LIKE HOLLOW SPHERE"

This was the headline of an article that appeared in the highly respected *Science News Letter* (April 22, 1961).

The opening statement sums up pretty well this conclusion:

"The Moon is like a hollow sphere, heavier on the outside than on the inside, according to the data from the Vanguard satellite and theories about the Moon."

This article's opening statement seems to be verified by the latest evidence—a peculiar characteristic of the Moon rocks Apollo astronauts brought back to Earth.

THE DENSITY OF MOON ROCKS INDICATES THE MOON MAY BE HOLLOW!

As we have seen, evidence and data scientists obtained on the density of the Moon long before our Space Age was launched indicated that our neighboring satellite was very light and possibly hollow. One of the best clues, therefore, as to whether this was a valid theory would come from the Moon rocks themselves.

In their six trips to Luna our astronauts brought back almost a short ton of lunar material—837 pounds, to be exact. I knew the density of Moon rocks would be a major clue to the make-up and density of the Moon itself. I searched and searched for accurate data on the lunar rocks, without success. I found no specific data on their density published in any of the NASA or lunar journals.

Then I got a revealing letter from another independent scientific researcher. He told me he had read my book *Our Mysterious Spaceship Moon* and was fascinated by the hollow-Moon theory.

> From reading it [the book] I gather that you are not aware of the most important fact about the moon's density. A fact that makes your theory more tenable. . . . After the moon rocks were analyzed, I looked repeatedly for data on their densities, but this was not published. I eventually called Dr. Paul Gast [one of NASA's leading scientists, who unfortunately passed away shortly thereafter] at his home and asked him what were their densities.
>
> He said they ran from 3.2 to 3.4. Now since the density of earth rocks averaged from 2.7 to 2.8 moon rocks were expected to be much lighter. You apparently saw the astronomers trying to drill about 30 inches with a drill that would easily go through 12 feet of earthrock (or steel supposedly) and the rocks obviously got harder and probably denser with depth. In fact, the authorities now say that the moon is differentiated, its heavier elements having settled, so one will have to assume that densities increase. How can they when the moon's density is no greater than the surface rocks unless the moon is hollow. . . .

This would verify the earlier data that the Moon is heavier on the outside than on the inside; as *Science News* puts it, "the moon is like a hollow sphere."

NASA VERIFICATION

Although specific data on the density of Moon rocks brought back to Earth were next to impossible to find, veteran science reporter Richard Lewis, who was among the NASA inside circle, got some information on the density of lunar rocks. He pointed out in his book *The Voyages of Apollo* (Quadrangle/New York Times Book Co., 1976) that lunar material brought back from Apollo 11 and 12 sites "were denser than Earth soils."

Again, how could this be? How could the Moon rocks be denser than Earth's (assuming of course that this general characteristic of denser lunar rocks held up in general for the entire Moon—and apparently it did), when other equally convincing and valid data indicate the Moon material, based on its global weight, is only about half as dense as that of Earth? Unless, again, the weight of the Moon is in its outer shell and the interior of the Moon is hollow!

A clear and distinct possibility and, considering all the other evidence, we think a clear probability!

THE MOON RINGS LIKE A HUGE HOLLOW SPHERE!

Another world of evidence exists to prove that the Moon is hollow—and it comes from inside the Moon globe itself.

Through scientific equipment and tests NASA and the world have learned a great deal about the Moon's interior. Scientific stations were set up on the Moon's surface which included sensitive seismometers that radioed a world of data back to Earth. The first were set up by Apollo 11 astronauts in the Sea of Tranquility, and another by Apollo 12 astronauts in the Sea of Storms. They were extremely sensitive—in fact, a hundred times more sensitive than any used on Earth, and able to record tremors almost one million times smaller than vibrations which human beings make. The sensitive seismometers even recorded the footfalls of our astronauts.

The first man-made crash directed at the Moon to divine its interior occurred after Apollo 12 astronauts had returned safely to their command ship and the lunar module ascent stage was sent smashing into the Moon's surface, thus producing in effect an artificial moonquake. The shock waves of this hit staggered NASA scientists—the Moon vibrated for over 55 minutes. Furthermore, the kinds of signals recorded by the seismometers were utterly different from any ever received before, starting with small waves, gaining in size to a peak, and then lasting for incredibly long periods of time. A wave took seven to eight minutes to reach the peak of impact energy and then gradually decreased in amplitude over a period that lasted almost an hour. It was claimed that even after an hour the minutest reverberations still had not completely stopped.

Amazingly, the LM hit the lunar surface about 40 miles from the landing site. The results were astonishing. All three seismometers in the package recorded the impact, which set up a sequence of reverberations lasting more than an hour. Nothing like this had ever been measured on Earth.

The impact occurred at 4:15 P.M. CST, November 20, 1969. A news conference had been scheduled to begin at 4:30 P.M. And when it did, strangely enough, the Moon was still "ringing" as the scientists—all of them seismic experimenters—arrived at the news center right out of their laboratory.

Maurice Ewing, co-head of the seismic experiment, told the afternoon news crowd of the stunning event, informing them that the Moon was still ringing. He confessed he was at a loss to explain why the Moon behaved so strangely.

"As for the meaning of it," Ewing explained, "I'd rather not make an interpretation right now. But it is as though one had struck a bell, say, in the belfry of a church a single blow and found that the reverberation from it continued thirty minutes."

Actually, unbeknown then to Ewing, the reverberations were to last for about twice that long.

Dr. Frank Press of the Massachusetts Institute of Technology was a bit more open and frank.

"We have been exploring for a hypothesis walking over here and usually when we speak too soon we are wrong. But let me say that none of us have seen anything like this on Earth. In all our experience, it is quite an extraordinary event. That this rather small impact . . . produced a signal which lasted thirty minutes is quite beyond the range of our experience. So, whatever it turns out to be, I think it will represent a major discovery, completely unanticipated about the Moon."

However, the remarkable thing is that this is exactly what would have been expected in light of the Vasin-Shcherbakov hollow-Moon theory. If the interior of this spaceship world really has a metallic-type hull and is really hollow, the seismic results of an Apollo spacecraft crashing into it should have produced different vibrations and tremors, probably of extremely long duration. The crash of a lunar module should make this metallic lunar ball vibrate like a huge bell. And it did!

Our own scientists were really at a loss to adequately explain the puzzle. As Richard Lewis puts it in his book *The Voyages of Apollo* "it was clear at the time of Apollo 12 that Apollo investigators had 'discovered' a new kind of planetary structure in the Moon. How it evolved, as well as the details of its nature, represented one of the most elegant mysteries of the twentieth century." (P. 124.)

Oh, they did attempt various explanations, which of course did not fit the seismic data.

Dr. Press of M.I.T. (now an advisor to President Carter) was one of the first to come up with some kind of explanation. He suggested that the LM impact might have touched off "a cascade of avalanches and collapses over a very large area."

This was not indicated, however, by the seismic data. For this indicated long, sustained response. The waves reached a peak between seven and eight minutes with hardly the slightest weakening evident at all. It took an amazing twenty minutes before the waves declined to half their strength. Then they gradually declined further, hanging on tenaciously for about an hour! This does not fit "the series of avalanches" explanation that Press offered.*

---

* The Apollo 11 seismometer left on the Moon in July 1969 had picked up about 100 similar seismic events which, though smaller, nevertheless had caused the Moon to resonate for up to twenty minutes. At first scientists were skeptical. As Dr. Gary Latham said: "No one was prepared to accept the fact that they might be

Another scientist suggested that the crash could have thrown lunar dust and debris so high into the atmosphere that it took almost an hour to come back down. Yet another proposed that "the LM itself might be to blame." This strained explanation held that the empty spacecraft perhaps did not hit straight down into the Moon but rather smashed like a crashing plane, bouncing across the Sea of Storms "like a flat stone on a stream, dropping off parts as it went."

However, as *Science News* (November 29, 1969) pointed out in dismissing all these weak explanations:

"Such a series of blows, however, would be unlikely to produce the sustained, large tremors recorded by the seismometer; the same argument applies to the falling-moon-debris theory."

RINGING THE LUNAR BELL

After the miracle of Apollo 12, scientists, particularly lunar seismologists, anxiously awaited the next big Moon crash. When Apollo 13's third stage was propelled out of Earth orbit into a Moon trajectory and by radio command sent crashing into the Moon, it hit with an impact equivalent to 11 tons of TNT. It hit about 87 miles from the site where the Apollo 12 seismometers were located.

The Moon gave a repeat performance—it shook, seismologically speaking, for more than 3 hours. In fact, 3 hours and 20 minutes, to be exact! Moreover, the vibrations of that artificially induced moonquake traveled to a depth of 22–25 miles. Scientists again were astounded. Again NASA seismologists were dumbfounded and could not come up with a satisfactory explanation.

But again, if the Soviet theory is correct, then such long reverberations are exactly what would be expected. Take any hollow, hard-crusted sphere, especially one that has a metal shell, strike it a hard blow, and it will behave exactly as our Moon behaved.

---

due to impacts." Then came the Apollo 12 impact, which caused the Moon to vibrate for nearly an hour. Comments Latham, "now we think differently."

THE MOON REACTS LIKE A HUGE GONG!

The next big blow to hit the Moon was that of Apollo 14's S-IVB, which was also boosted into a lunar crash course and by remote control sent smashing into the Moon's surface. The moon behaved predictably, in the same way.

As a NASA science publication (*Apollo 14: Science at Fra Mauro,* 1971, p. 17) tells us: *"The Moon reacted like a gong. For about three hours it vibrated and these vibrations traveled to a depth of from 22–25 miles."* (Emphasis added.)

This despite the fact that the vehicle crashed more than 108 miles away from the Apollo 14 instrument site!

Similarly, the crash of the lunar landing craft after the Apollo 14 astronauts had returned safely to the command module *Kitty Hawk* caused the Moon's bell to be rung. The empty ascent stage, which weighed 4850 pounds, hit with a force equal to the explosion of 1600 pounds of TNT. This time the tremors reverberated around the Moon for 90 minutes.

This NASA publication says: "This minor moonquake marked a scientific milestone. It was the first time any event, manmade or natural, was recorded by a network of two seismic stations on the Moon."

Later observations (made possible when three different seismic stations on the Moon recorded a man-made impact —Apollo 12, Apollo 14, and another left by Apollo 15 astronauts at the Hadley-Apennine landing site) indicated that seismic signals traveled over 700 miles from the Apollo 15 impact site to a seismometer on the Sea of Storms, that great distance away, and then all the way to the seismometers at the Fra Mauro highlands. Scientists admit that on Earth similar signals produced by such an impact probably would not have traveled more than a few miles at the most. Nor would they have lasted more than a minute or so at the most. Certainly not for hours.

Unquestionably, the Earth's interior, which is solid rock, is different from the Moon's, which seems to be more like a hollow sphere. Dr. Latham, principal investigator of the seismic experiment, admits that the Moon's strange behavior continued to baffle scientists, who have found no counterpart on our planet Earth. Apparently this is because the Moon is so different interiorly.

In fact, scientists readily admitted that seismic evidence does strongly point to the fact that the Moon is rigid and cold on the interior, with no molten lava core. On Earth similar seismic waves are damped out and absorbed by the planet's great mass and by the Earth's molten core. Scientists know, however, that a hollow Moon—either completely hollow or with just an inner hollow layer all the way around—would behave as the Moon *does*.

All man-made crashes into the lunar surface producing moonquakes ended up with similar results. Indeed, after one impact the Moon rang for over four hours!

However, impressive as these hits were, they do not prove in the orthodox scientist's mind that the Moon is completely hollow. At least across the center. Other seismic tests have produced peculiar results, leading a few to believe that the Moon may be solid to the center. However, if the Moon is a spaceship, then, as expected, it would have all kinds of artificial structures on the inside, including, as we shall see, two seismic belts that at least one leading scientist from NASA says could be two huge girderlike blocks of metal 1000 kilometers long! Would not such constructions foul up any conclusions about whether the Moon is solid or has a core?

One problem is that the present seismometers on the Moon were placed too close together. If they were farther apart they would be much more effective in accomplishing the feat of proving once and for all whether the Moon is hollow. As one astronomer from a Midwest observatory told a radio audience, discussing my book *Our Mysterious Spaceship Moon,* our satellite might very well be hollow. However, the evidence from present seismic equipment is unfortunately not quite clear or conclusive, simply because they are too close together to get a true readout between the primary (p) and secondary (s) waves.

If the Moon is completely hollow across the center then the primary waves would not cut completely across the center of the Moon, but the secondary waves would race around the outer shell. The time difference could establish the elusive proof that the Moon is (or is not) hollow.

The answer to this question, of course, hinges on another key question: Does the Moon have any core? Science reporter Richard Lewis said that "Urey and others insisted that the Moon could not have a core because of its lower density, but some geophysicists were not convinced." (*The Voyages of Apollo*.)

Scientists were hoping to have proof once and for all through the impact of a large meteor whose reflected waves would indicate whether a core did or did not exist.

Fortunately, scientists got a once-in-a-million-years miracle when a large meteor, which Apollo scientists dubbed "a whopper," slammed into the Moon on May 13, 1972, with the energy equivalent of about 200 tons of TNT. But, surprisingly, although the impact sent seismic shock waves down through the crust and well into the Moon's inner layers, the "whopper" did not send any reflective waves back.

Dr. Latham, NASA's director of the seismic project, commenting on this unusual turn of events, said: "We have been searching for a signal—energy which would have gone down to the core, and bounced back up as a reflector." The whopper should have gone down to the lunar core and bounced back up, but it did not. Scientists wondered why it didn't. Could it be another indication that, as Dr. Urey suspects, the Moon has no core? That it is in fact, as the Soviet scientists tell us, *hollow?* (*Science News*, July 1972.)

Dr. Urey does tell us that "transverse seismic waves are not reflected back because it is felt that the soft structure of the middle of the Moon absorbed them." However, as we have seen, there is much evidence that the Moon is rigid at great depths. Could it not be that these shock waves are not reflected back simply because the Moon is wholly or at least in part hollow? Or are our scientists having great difficulties divining the inside of the Moon because it is not entirely natural? Irving Michelson pointed out in *Man On the Moon* (Eugene Rabanivich, ed. Basic Books, 1969) that "it is a curious fact that although the moon's rotational motion has been very accurately known for nearly three centuries, attempts to infer values of the inertia moments have been notoriously unsuccessful." Couldn't this be the reason?

It should also be remembered that if the Moon really is

a spaceship scientists must expect all kinds of artificial construction on the interior. (Later we shall see there is evidence of this.) Hence, the nature of the reflective waves and seismic waves in general would be very confusing. And, as we have seen, that has been exactly the case.

For instance, an early motion study indicated that the Moon was hollow. The coefficient of moment of inertia, which is critical for understanding the density distribution of the lunar interior, yielded an early value (0.6) implying a hollow Moon. Subsequent studies have conflicted with this result. But again we point out that an unknown variety of internal constructions could be upsetting the correct data. This should be taken into consideration in arriving at any correct conclusion about the true nature of the Moon.

For many studies indicate that the Moon is hollow. Its extremely low density indicated it might be. H. P. Wilkins claimed it was, at least to a large measure. The early MacDonald motion studies and the later Solomon gravitational-field studies indicated it might be. Impact shock waves which vibrated through the Moon for up to four hours indicate that it is hollow. So though the problem remains unresolved, any further studies should take the Vasin-Shcherbakov hollow-Moon theory and possible artificial spaceship construction into consideration in arriving at a true picture about our Moon.

Amazingly, there is some indirect evidence that our space agency officials are taking the "hollow" problem of the Moon more seriously than they are letting the public know. For it is interesting and, we believe, important to note that Dr. Farouk El Baz, one of NASA's leading scientists before he went to the Smithsonian Institute as director of space research, claims that NASA suspected hollows did indeed exist within the Moon and admits that our government ordered and carried out certain experiments on the Moon to determine if they did. He claims that such experiments, *which were not publicly announced*, were held in the strictest secrecy. Says El Baz: "There are many undiscovered caverns suspected to exist beneath the surface of the Moon. Several experiments have been flown to the Moon to see if there actually were such caverns." (*Saga*, March 1974, p. 36.)

No results have ever been announced. In fact, even the

experiment was kept secret—at least until El Baz made his public statement in an interview with *Saga* magazine, one of America's leading journals, which is in the forefront of UFO investigation, and has also been effective in penetrating our government's secrecy in the space field.

Why all this secrecy? Is it because our government and its space agency, along with our military, consider (contrary to public knowledge) that hollow areas exist inside our Moon—that possibly alien bases, as some suspect, exist in these underground Moon hollows? Did alien beings take advantage of these hollows to create, as Soviet scientists speculate, millions more square miles? Or did they perhaps hollow out the central area of the Moon completely?

In the book *Man On the Moon* Irving Michelson points out that "from studies of the translational motion of the moon" (i.e., motion of its center of mass) there exists evidence of "the non-uniformity of the internal mass distribution within the moon," which indicates that "the densest regions of the moon are nearest to its surface, in what has been termed the 'hollow moon' hypothesis." But, objects Michelson, "the apparent density inversion needed to provide consistency in the theory of the moon's orbital motion, termed the lunar theory, is not only unfamiliar but wholly unacceptable on a physical-mechanical basis—a hollow shell of this type would necessarily collapse."

In answer to Michelson's objections, naturally it is "unfamiliar," for it is not natural—we are dealing with a world that is not completely natural but partially artificial. Of course, it is unacceptable on a natural physical basis. But this is something that apparently orthodox scientists never thought of (until our Soviet theorists, that is): that we may be dealing with a hybrid world—part natural and part artificial.

As to the last objection, that a hollow Moon is "wholly unacceptable . . . [because] a hollow shell of this type would necessarily collapse," we point out that the Moon, as we shall see, has an extremely strong metallic inner shell—the hull of a spaceship! And so Michelson's objections fall.

And so the evidence of a hollow Moon stands!

The surface of the Moon is a dark gray, gunmetal gray. It looks like molten lead that has been shot with BBs.

—Astronaut James Irwin, Apollo 15

NINE
DOES THE MOON HAVE A CONSTRUCTED OUTER SHELL?

• What are the strange circular dark areas of the Moon and why do they *not* seem to be placed randomly across the surface of the Moon?

• Why is the far side of the Moon so different from the near, or Earth, side? Why are the strange lunar seas virtually nonexistent on the Moon's far side?

• Why is the Moon so pockmarked with craters and, more significantly, why are they so uniformly shallow? Why do two Soviet scientists think that this indicates that the Moon has an artificially constructed outer shell?

• Why are the maria so loaded with metals and metallic elements—titanium, yttrium, and other refractory metals which require high temperatures to melt?

• Why do Soviet researchers reach such a drastic conclusion that the Moon's maria are in fact the evidence of the hollowing-out and reinforcing process in the reconversion of the natural world into a huge spaceship by unknown alien beings?

• Why do our scientists find it extremely difficult to account for the high amount of metals like iron and titanium in the Moon's dark areas?

• How do the scientists account for the fact that the Moon's outer surface somehow reached more than 4000°F.?

• Why do scientists even today find it impossible to account for the formation of the maria? Why do all the theories have fatal flaws?

• Why do some scientists suggest that the Moon's maria were formed by a special kind of volcanic action that scientists of Earth are not familiar with?

• Why does the Moon's surface give evidence of formation by artificially induced volcanic action?

• How do scientists account for the pure metallic elements found in lunar samples (such as pure titanium, used on Earth in the manufacture of spacecraft and supersonic jets), which seem to be in such great abundance in the maria?

• How do scientists account for the shocking and unbelievable discovery of pure metallic particles in lunar samples? How do they explain the presence of pure iron particles, especially pure iron that does not rust? Why does this indicate a manufacturing process from a technology well beyond man's?

• How do scientists account for the fact that they cannot reproduce lunar materials even in our most advanced scientific labs?

• Why are all the maria so metal-rich?

• How do scientists account for the fact that on the Moon denser materials are on the surface? How do they explain how heavier materials can flow to the top of the Moon? Is this not seemingly impossible unless artificial construction methods were used?

• Why does the Moon have such a thick outer shell, melted down to great depths? How does one get a 65-kilometer-thick crust without melting most of the Moon? And if melting occurred—as evidence indicates it did—then how could the Moon's interior today be so relatively cool?

• How does the Moon give indications that "some force or forces" rearranged the Moon's outer surface? What force brought about the significant redistribution of the Moon's crustal interior?

Anyone who peers at the Moon through a telescope is impressed at how different this satellite world is from the lush verdurous and water-covered Earth. The barren Moon appears to be dominated by two rugged features: crater and maria, those strangely level, circular dark areas that Earth-bound man once mistakenly thought were oceans of water. Though we now know that these dark, strangely level regions appear to be oceans of lava, man is still at a loss to adequately explain how they came into existence or even what they are. Even before we went to the Moon some guessed that these weirdly dark splotches, which remind one of freshly cleaned blackboards, are actually loaded with heavy dark mineral ores, such as iron and titanium.

Our sea-laden world is three-fourths covered with water. The Moon is covered with seas of apparently once molten rock. The question is, where did all the lava come from that created these immense oceans of lava covering a third of the Moon's near side? An extremely hot interior? Or from impaction of huge meteors or asteroids?

Still another mystery lurks here, raised by the Moon probes—that of the differing far side of the Moon, which has no extensive maria.

THE MOON'S HIDDEN FACE

Not once within mankind's memory has the far side of our Moon shown its face to Earth, and most scientists assumed it would be no different from the side that is familiar to us. While it is true that a few thought it would be radically different, it is clear that they were speculating wildly. And some were really wild.

For instance, Hanson, a Danish astronomer of the eighteenth century, came to believe that the far side was inhabited. In studying the movements of the Moon, he had found one tiny discrepancy that he could not explain: The Moon did not appear uniform in density to him. To explain this Hanson theorized that one hemisphere of the Moon was slightly more massive than the other. He concluded that all the atmosphere of the lunar world (which he insisted the Moon had), as well as all its oceans and seas, had been drawn to the far, hidden side. He also was convinced that this side could very well be inhabited and probably teemed with cities.

Hanson was, of course, mercilessly ridiculed by his contemporaries for his way-out notions. One anonymous English poet even penned some fancy lines in his honor:

> Oh Moon, lovely Moon with the beautiful face,
> Careering throughout the bound'ries of space,
> Wherever I see you, I think in my mind—
> Shall I ever, O ever, behold your behind?

Most scientists still remained convinced that the far side of the Moon was no different from the side they had studied over the centuries. Then came the Soviet space probes which first photographed it, and the world was amazed to discover how different the far side was. No serious scientist had expected to find any cities or Moonmen, but what piqued their curiosity and raised several more tantalizing questions was how startlingly different the back side really was. It was far more rugged and far more pockmarked with craters and riddled with mountains. And, strangely enough, it had few maria, or seas of lava.

Why should these massive, dark plains be so prevalent on the Earth side and so lacking on the far side? This strange dichotomy has mystified scientists since man began to extensively see and photograph it. The mystery still baffles scientists today.

The inference, of course, was that these huge, circular seas could not have been formed by chance meteor and asteroid impaction; these bodies would necessarily have hit at random all over the Moon. The evidence seemed instead to indicate that the maria on the near side were formed by a welling up of volcanically produced lava from deep within the Moon. And here the problems really began.

Another conundrum was the fact that the far side has far more rugged and mountainous exterior, with many more craters. And this brings us back to the mystery of the craters themselves, their strange shallowness and the problem of just how they were formed.

Most scientists are convinced that lunar craters were blasted out by celestial bombs, all right, hitting with the force of many tons of explosives. But they were of the atomic variety in explosive equivalency—huge meteors, asteroids, or comets. The explosive force of a huge meteor smashing into the Moon at speeds of up to 30,000 miles per second would be equivalent to millions of tons of TNT —actually dwarfing to insignificance the tiny atomic blast that left Hiroshima a desolate crater. Undoubtedly in some cases the powerful explosions on the Moon were hundreds of times greater than the first bombs detonated on Earth.

It has been calculated by Soviet scientists that a meteor of one million tons is equal in explosive force to a single megaton atomic bomb, which is about fifty times the Hiroshima size. Yet we know that gigantic meteors weighing one million tons have clobbered the lunar surface, leaving only broad vast shallow holes in the Moon's skin.

Why this should be has been a puzzle to astronomers. Vasin and Shcherbakov suspected that the answer to this mystery lay in the Moon itself. The consistency of shallowness of Moon craters puzzled them. Despite their size— some of them fantastically large by Earth standards—almost without exception even the greatest gaping holes are surprisingly shallow. Even craters 100 miles or more across appear to be no deeper than a mile or two. Why?

What could be the reason for this strange phenomenon? Certainly meteors 50 to 100 miles in diameter blasting into the Moon and exploding with the force of millions of atomic bombs should have torn deep holes in its surface. The terrific impact of a meteor dozens of miles across colliding with our Moon should have produced an explosion so fierce that holes many miles deep would have been gouged out. But nowhere has any celestial missile done this.

Even conservative estimates of expected results indicate that meteors 10 miles or more in diameter—larger than most that have hit Earth—should have penetrated to a depth equal to four to five times its diameter. Yet the deepest crater we know of on the Moon is the Gagarin Crater, which, though 186 miles across, is less than 4 miles down. Most are like Clavius, which is a whopping 146 miles across but only 2–3 miles deep, and this despite the fact that all of Switzerland and Luxembourg could be placed inside it, it is so large. There is no question that whatever hit here should have gouged a great hole scores of miles deep. Why didn't it?

OUR MYSTERIOUS "LOP-SIDED" MOON WITH ITS INTERIOR IRON-RICH LAYER

SCHEMATIC CROSS-SECTION OF THE MOON

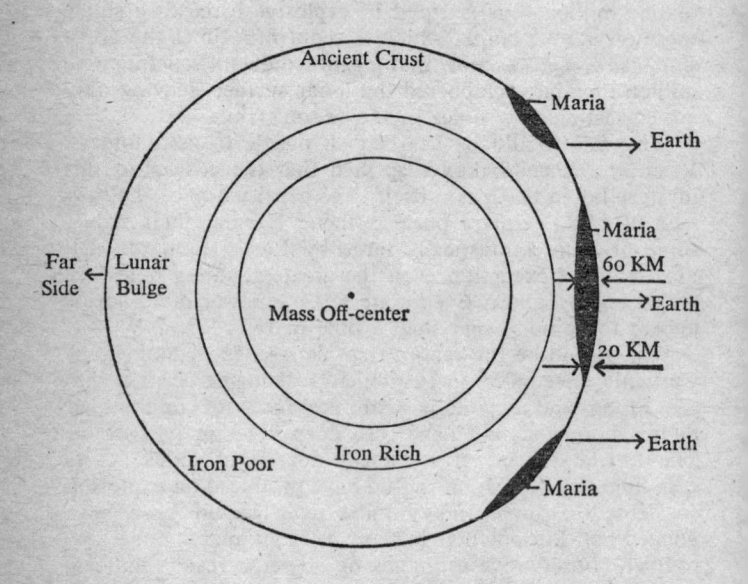

Shows: (1) How moon is lop-sided
(2) Iron-rich inner layer—
Could this be reinforced inner hull made from metallic-rich rock?

Adapted from *Physics Today*, March 1974.

The answer according to some scientists lies in the fact that most craters on the Moon were not caused by celestial bombardment but are volcanic in origin. However, the vast majority of lunar scientists having studied lunar crater characteristics held forth the impact formation of the origin of craters. Dr. Harold Urey, one of the world's most knowledgeable lunar experts, stated the majority opinion in 1956 when he claimed that the vast majority were caused by celestial crashes.

A SOLUTION TO THE SHALLOW-CRATER PROBLEM?

Our Soviet spaceship speculators suggest that the answer to the shallow-crater mystery lies in the Moon's make-up. The shallow craters, Vasin and Shcherbakov insist, would perhaps be strange to a normal satellite, but the Moon is a "remade" world with an inner and outer shell of "armour plating."

They note: "if one assumes that when the meteorite strikes the outer covering of the Moon, this plays the role of a buffer and the foreign body finds itself up against an impenetrable spherical barrier. Only slightly denting the 20 mile layer of armour plating, the explosion flings bits of its 'coating' far and wide.

"Bearing in mind that the Moon's defense coating is, according to our calculations, 2.5 miles thick, one sees that this is approximately the maximum depth of the craters."

Much of this outer portion of the Moon's armor protection, these spaceship theorists point out, is furnished by the maria—those strange huge dark areas of filled-in metallic rock that can be seen with the naked eye on any clear night that the Moon is visible in Earth skies.

These huge circular seas or maria help form the familiar Man in the Moon which many Earthlings have come to see in our companion world. On our very first trip to the Moon we landed in the left eye of the Man in the Moon—the Sea of Tranquility. The Sea of Rains makes up the right eye.

Some of these strange dark seas are immense in size.

The Sea of Storms, for instance, is over 2,000,000 square miles, which makes it larger than the entire Mediterranean Sea! The Sea of Serenity, on which we landed our Apollo spacecraft, is about the size of Great Britain and France put together. And considering that the Moon is a much smaller world than Earth—about one fourth the size—makes them much more impressive in size.

Another maria, the Sea of Rains, embraces an area that in the United States would extend in one direction from Michigan to North Carolina and from South Carolina and Georgia to Arkansas and Iowa in the other. This vast sea, which is about 750 miles across, covers more than 340,000 square miles. Before we went to the Moon most scientists speculated that this immense sea was produced when an asteroid the size of the island of Cyprus collided with the Moon and exploded there.

Although some of the people in the past thought these dark areas of the Moon were huge seas, most modern scientists felt that they were in fact huge seas of lava. Others were skeptical of this theory that they were formed from immense lava flows which poured out of the Moon's extremely hot interior when the lunar globe was very young. Where could all the lava come from, many objected, to create such immense oceans of lava that cover one third of the Moon on the Earth side? Many scientists did not believe that the Moon ever had gotten that hot inside, maintaining that the Moon was formed cold. Why did they hold this? Because they calculated the Moon was too tiny a world to generate the kind of heat necessary to produce such vast lava flows. Some of these scientists were convinced that the answer to the formation of the maria lay in the celestial missiles that bombarded the planetoid—meteors, comets, and asteroids which, exploding upon contact, threw vast quantities of liquid lava over the face of this world. Yet there existed strong evidence that maria were formed by lava welling up from its interior. Moreover, the maria do not appear to be randomly scattered across the Moon's face; most of them lie in one quadrant of the Moon's Earth side. This militated against the random-peppering theory of impactists.

And this brings us to yet another mystery of the maria. For another peculiarity of these peculiar lunar features is their distribution. Many are circular in form, conforming strangely enough to the general spherical shape of this mystery world. Their distribution was also unusual—almost all of the Moon's maria on the Earth side of the Moon and four fifths of them on the far side lie in one quadrant—in the right-hand section or area of the Moon.

Scientists on Earth were still arguing how the maria on the Moon got there when our astronauts landed. They were in for some shocking surprises, uncovering evidence that indicated to the two Soviet scientists that the maria were not natural formations.

Shcherbakov and Vasin became convinced from this evidence that the maria were in fact created by artificial means, probably by "Moon beings" pouring out huge inner portions of lunar lava and metal not only to hollow out portions of their inner world but to reinforce their metallic armor and prevent it from being damaged by the "impact of celestial torpedoes" crashing into it from time to time.

Vasin and Shcherbakov wrote: "The Moon's dry seas are in fact areas from which the protective coating was torn from the armour cladding. To make good the damage to these vast tracts . . . [they flooded] the areas with its 'cement' resulting in flat stretches that look like seas to the terrestrial observer.

"The Moon's population presumably took the necessary steps to remedy the effects of meteorite bombardment, for example, patching up rents in the outer shield covering the lunar shell. For such purposes a substance from the lunar core was probably used, a kind of cement being made from it. After processing this would be piped to the surface sites where it was required."

WHY THE MARIA ARE SO LEVEL

If these spaceship theorists are right then a number of mysteries of the Moon's maria would be resolved—for instance, how they were formed and why they have such seeming nonrandom distribution. Another puzzlement would then become understandable too—the strange levelness of the seas. For as science reporter Henry Cooper points out in his book *Apollo on the Moon* (Dial, 1969), one of the questions that troubled NASA scientists was the answer to the mystery of "how were the maria leveled so neatly?" Artificial formation by alien intelligence would, of course, help make understandable what Cooper refers to as the "billiard-table standards of the maria."

METAL IN THE MOON'S MARIA?

What evidence led these two Soviet scientists to such a dramatic conclusion—that the maria are in fact the evidence of the hollowing out and repairing of a huge spaceship world?

The evidence they offer is fragmentary, based almost entirely on the fact that Moon rocks contained large amounts of metallic elements—titanium, chromium, and zirconium, highly refractory metals which are not only heat-resistant but mechanically strong and have many admirable anticorrosive properties. Say Shcherbakov and Vasin; "A combination of them all would have enviable resistance to heat and the ability to stand up to the means of aggression, and could be used on Earth for linings for electrical furnaces."

In fact, the indisputable evidence that great amounts of such metals were found in Moon rocks, though impressive, is about the only evidence they offered for this bizarre, radical thesis. Admittedly, such a finding is difficult to "explain away," but was any other evidence uncovered by America's Moon probes that would support this wild theory?

What did we find when we went to the Moon? First of all, the Moon rocks that we brought back for analysis were rather puzzling. Dr. Eugene Shoemaker of the California Institute of Technology, chief geological interpreter of the Apollo results for NASA, confessed that the Moon samples raised "ten times as many good questions as they are likely to answer."

Our first explorers brought back samples from the mysterious Sea of Tranquility. The rocks from here astounded the scientists who so eagerly and carefully examined them. It was found that they—and presumably the solidified lava "seas" from which they were taken—had heavy concentrations of tough, high-heat-resistant elements like titanium, which indicated to scientists that the Moon's outer layers were melted by some source of heat that was very intense.

NASA scientists estimate that parts of the Moon's outer shell must have reached at least 4000°C. to melt and fuse with rock the kinds of elements found there. The perplexing problem is this: How could the surface of the Moon have reached such extremely high temperatures? Scientists frankly were at a loss to adequately answer this question.

Some of the Apollo samples proved to be ten times as rich in titanium as the most titanium-rich rocks ever found on our planet Earth. But not only titanium was found in these lunar samples, but also zirconium, yttrium, beryllium, and similar rare, anticorrosive, temperature-resistant metallic elements in perplexingly high percentages. These are among the most durable elements known to man and have admirable heat-resistant qualities, requiring extremely high temperatures to melt. Yet lunar samples indicate that they are present in shockingly great abundance in the lunar outer crust. In fact, one scientific journal (*Science News*, August 16, 1969) states that the outer surface of the Moon seems to contain titanium, yttrium, and zirconium in "amounts higher than present estimates either in earthly rock or estimates of elemental abundance in the universe."

Dr. Harold Urey confessed freely to his colleagues that he "was terribly puzzled by the rocks from the moon, and in particular by their titanium content." To him this was a "mind-blowing" discovery. He admitted candidly: "I just don't know how to account for the titanium." (Henry Cooper, *Moon Rocks*, Dial, 1970.)

But Urey was not alone. For not a single scientist could solve the conundrum. To make matters worse, many scientists, like Urey, had concluded before the Moon expeditions that our satellite was essentially a cold body, that it had never been hot enough for any significant lava flows because it just was not large enough to produce enough heat to cause lava flows to any great extent. Now scientists had to figure out how such a small world could have generated temperatures as the data indicated it once had. It was most embarrassing for Urey, because before Apollo he had insisted that he could prove mathematically that the Moon was too small an orb to generate the kind of heat necessary to produce significant volcanic lava flows.

Another respected scientist, Dr. S. Ross Taylor, the geochemist in charge of chemical analysis, pointed out: "The problem is that on the Moon maria the size of Texas had to be covered with melted rock containing fluid titanium and one wouldn't expect titanium ever to be hot enough or viscous enough to do that even on earth—and no one had ever suggested that the Moon was hotter than earth." (*Moon Rocks.*)

What could have produced heat of such intensity? Scientists do not know how it could have been done *naturally*. And here is the key. For all orthodox scientific considerations fall short of the mark in solving this problem. However, if scientists accept the unorthodox "spaceship" view of our two Soviet scientists that this intense heat was used to artificially process great amounts of refractory metals to flood the outer areas of the Moon, then the problem disappears.

But need we necessarily turn to artificial construction by alien intelligence to solve the problem? No, for there is another theory, common even before we went to the Moon, which some scientists thought answered the difficulties.

WERE THE DARK LUNAR PLAINS PRODUCED BY IMPACT MELTING?

Long before man journeyed to this mystery world, scientists had wondered how the Moon could have generated the kind of heat necessary to pour forth oceans of lava and cover one third of the surface area of this world. Now the problem suddenly became much more complex, for evidence indicated that tremendous amounts of heat-resistant metallic elements existed in the maria. How then could scientists account for this fantastic heat, which seemingly eliminated natural volcanic action? Some claimed that it could have been done by celestial bombardment of gigantic meteors, asteroids, or even comets.

One leading scientist who came to this conviction was Zdenak Kopal of the Department of Astronomy at the

University of Manchester. This leading lunar light examined the early Apollo evidence and in the prestigious publication *The Moon, A Physics and Astronomy of the Moon* (Academic Press, 2nd Ed., 1971 edition) insisted that the great plains of the Moon could hardly have been lava flows from the deep interior but were probably produced instead by the tremendous energy generated by great cosmic collisions of meteors and the like, perhaps along with some heat produced by radioactive elements concentrated in the near surface areas.

Is this the solution to the seeming insoluble problems of the Moon's maria? As we have seen, the impact theory has its weaknesses too. For one thing, it is hard to conceive how one third of the Moon's near side just happened to be hit that hard by random celestial bombardment while the far side, though presumably subject to the same bombardment, was hardly touched. As we have seen, the far side has hardly any maria areas, while the near side is full of them. Furthermore, if radioactivity were involved in the heat-producing process, the far side with its much thicker crust should contain much more of the radioactive elements that helped to produce lunar lava flows. Yet it did not— they are strangely absent there.

Also, evidence from the Moon's maria itself indicates that these huge circular seas of lava were formed from an upwelling of volcanically produced material generated deep inside the Moon.

Nevertheless, the impacting of celestial missiles as the cause of maria did seem to be the only answer to this problem.

However, in 1976 the great lunar expert Kopal published his book *The Moon in the Post Apollo Era* (Reidel, 1974), updating with the latest lunar evidence and findings what happened to our Moon. Not surprisingly, Kopal completely changed his mind, now discounting the possibility that the Moon's dark circular plains were caused by celestial impactings. Kopal confesses, "it was clearly not the impact heat which melted the magmas now covering the mare floors; for the latter must have exuded from the interior at a considerable later time." The dating of Moon rocks and other evidence had led him to this conclusion. He adds: "A quest for the source of the basaltic magma we now see spread over the surface of the mare basins should, therefore, lead us back again into the lunar interior."

Kopal points out that immense difficulties "are encountered, however, when we consider the mechanism, which could have driven out such lavas to the surface. . . ." He shows that the structure of rocks taken from these mare areas indicate that they were differentiated geologically, cooked under pressures that indicate that they came from depths of not less than 150 kilometers. This is deeper by far than "the depth of volcanic chambers" on our planet Earth. Kopal freely admits the immense difficulties of getting lava from such great depths inside our Moon to the surface, saying that "to pump molten material from such depths up the surface calls for an expenditure of energy whose source is not easily apparent." (*The Moon in the Post Apollo Era.*)

So not only can't scientists find out how the Moon could have generated the kind of intense heat necessary to melt high-temperature metals like titanium, amalgamating it with rock, but they cannot figure out what force or energy could have pumped such vast oceans of lava to the surface, spewing it forth in veritable seas across the outer portions of the lunar planetoid.

STILL MORE PUZZLING PROBLEMS WITH THE VOLCANIC SOLUTION

Scientists today continue to wonder where the energy for such large circular seas of lava could have come from. Did all strangely dark lava come from the natural volcanic action of an internally hot Moon? Dr. Gerald Wasserburg of the California Institute of Technology points out this presents other serious problems in light of the established fact that our Moon is a relatively cold body today. Before scientists can accept this theory, as Wasserburg points out, "we must attempt to understand what the precise mechanism is by which the thermal energies of the Moon were shut down to prevent further volcanism." Wasserburg indicates that we have here another "impossible" situation.

One of the major theories as to how the Moon could have accumulated enough heat to produce these huge seas of lava suggests the accumulation of radioactive elements in the surface areas, generating enough heat to melt internal sections of the Moon and pour forth these then-liquid lava oceans to the surface. But as the journal *Scientific American* points out, this too has serious problems, for the heat produced by radioactive material found in abundance on the Moon is slow and gradual. Evidence indicates that the Moon melted rapidly so that the lava became "very fluid and flowed over these areas quickly." So leading Moon experts hold that other methods must be thought of.

But what other solutions are there? Lunar experts themselves have offered compelling reasons why impact melting is out. Now apparently the only other solution, internal volcanic action, cannot be the answer either. Which way shall scientists turn? To the Soviet spaceship theory? Should we point out that all problems, puzzles, and conflicts disappear in light of the Soviet artificial-Moon theory?

ANOTHER KIND OF VOLCANISM!

In the midst of all these contradictions and conflicts one NASA scientist suggested that maybe it was all so confusing because scientists were looking at "a special kind of volcanism."

Exactly! It is now clear that what took place on the Moon is a volcanism artificially induced, using (as we shall see) some kind of radioactive melting done in an artificially created way, whereby aliens might have mixed great amounts of metallic elements from the Moon's deep interior with molten rock. For this is no ordinary rock, as assayed lunar samples have proven. It is full of metals.

No wonder—and this television watchers of these historical expeditions may well recall—our astronauts, trying to drill through the maria, found to their frustration and scientists' amazement that they could hardly penetrate the tough peculiar skin of this world. This despite the fact that they had specially designed drills that could go through just about anything. On several occasions the astronauts using all their strength could hardly penetrate this tough material, and for all their efforts merely got down a few inches! Here again is more evidence that this is not a natural world. But this is just the beginning of such evidence. For the mind-boggling constructions on the Moon were to yield amazing evidence that these are just what they seem to be: constructs!

PURE PROCESSED METALS ON THE MOON?

Admittedly, the artificial-Moon theory is the easy answer—almost the obvious answer—to the enigmas and problems offered by the Moon. But is there any hard evidence to prove this far-out theory, other than the fact that it dissolves the immense difficulties of this difficult-to-understand world? Just because there seems to be no natural way to explain how this complex companion of our Earth came into being is not sufficient ground to label it a spaceship. Surprisingly, such evidence has been unearthed (or should we say "unmooned"?).

It has been definitely established, as we have seen, that these strangely level, dark lunar plains are loaded with dark minerals like titanium, a metallic element used on Earth in the manufacture of supersonic aircraft and even spacecraft.

This startling find, however, is only the beginning!

For scientists scrutinizing the precious treasure of 837 pounds of the Moon (and undoubtedly it is the most carefully, meticulously, and thoroughly examined and analyzed group of rocks man has ever looked at) have discovered, much to their amazement, actual pure metal particles in the lunar samples!

General Electric Research and Development Center issued a scientific report declaring that pure iron particles have been found in lunar samples. When this startling report first came out it caused many scientists to question the validity of GE's findings. For on the surface it would seem to be scientifically impossible. Pure iron particles found in nature? However, the findings have been since verified by the University of California (Berkeley), proving beyond a doubt the presence of pure iron particles in lunar samples. (*New York Times,* January 7, 1970.)

Could they not, however, have come from meteorites? No, insist the scientific experts. John Noble Wilford, science editor of the *New York Times,* points out: "The moon's iron particles apparently did not come from meteorites because the iron in meteorites occurs in an alloy with nickel." (*New York Times,* January 7, 1970.)

Miraculously, Soviet remote-controlled, unmanned lunar probes (Zond 16–20) which have brought back tiny samples of the Moon also found pure iron particles. In August 1976 Soviet scientists brought back more treasures from the Moon—a soil sample from the Sea of Crisis. It was at this time that the Soviet government announced that their scientists had discovered pure iron particles in their lunar samples. They added something very important in their report: *These lunar iron particles contain iron that does not rust!*

The Associated Press wire carried this brief but startling Soviet announcement:

"Emphasizing the importance of lunar soil samples, an article in *Pravda* [official Soviet government news journal] revealed that the first successful automatic mission in 1970 brought back particles of iron that 'does not rust.'" Pure iron that does not rust is unknown on Earth. In fact, it cannot yet be even manufactured. Physicists and scientific experts claim they cannot understand how this is at all possible without some kind of manufacturing process being involved. They also point out that it is beyond our present Earth technology. Pure iron that does not rust is not found anywhere in a natural state. (*Detroit Free Press,* August 24, 1976.)

Unfortunately, the *Pravda* story passed over the staggering implications in this discovery of pure iron that does not rust. *Pravda* merely comments:

"If we understand how such iron is formed on the Moon and we can learn to manufacture it under earth conditions, this [discovery] would repay all the expenditures for space study," according to a Soviet scientist.

But this is not the end of the amazing story of metals found on the Moon. Cambridge University scientists have also found brass, mica, and amphibole in lunar material. And beyond this, the near-pure titanium that was found there stunned lunar experts.

Scientists now know that it is the extraordinarily high amount of titanium which helps make the maria dark. The mineral in which titanium resides (illeminite) is black and opaque, and it imparts a dark color to the lunar lava. This black mineral is why these areas look so dark to man on Earth.

Scientists have been doing lunar flip-flops trying to explain the existence of pure titanium on the Moon. *Science News* carried one of these strained attempts. (January 17, 1970.) The theory is that titanium darkens as it becomes "closer to pure metals." Scientists theorize that maybe the solar wind is responsible for this wonder of wonders, perhaps by knocking atoms loose from their oxides, causing minerals such as titanium "to darken as they become closer to pure metals."

This is one scientific speculation which tries to account for the discovery of near-pure metals on the Moon—so pure that they seem almost to have been manufactured by some tremendously advanced technology. Even if we accept this farfetched speculation—and the scientists who devised it admit themselves that it is just that—it still does not solve the other problems, such as how such intense heat could have been generated on this relatively tiny cold orb in our skies.

SCIENTISTS ON EARTH CANNOT EVEN REPRODUCE LUNAR MATERIALS!

The entire package of lunar samples from the maria—which were apparently cooked in extremely high temperatures—has left scientists almost totally perplexed. They have studied these lunar samples minutely and as yet they are completely stumped to explain them.

Dr. Robert Jastrow points this out in his article "The Moon is a Rosetta Stone." (*New York Times Magazine*, November 1969.) Here is another lunar impasse: "No pattern emerges from the discrepancies. The geochemists cannot see any way in which the materials of the Earth, even if worked over repeatedly by long-continued sequences of melting and remelting and recrystallization, could be made to yield the detailed composition of the Apollo 11 rocks." Like the pure iron which does not rust, they almost seemed to have been manufactured—produced by a technology far beyond our own.

THE METALLIC-LOOKING MARIA

Many areas of the Moon seem to be covered with metals, as astronaut James Irwin observed (see quote heading this chapter). In an ancient issue of the *Scientific American* (February 1927), following up an article on the peculiar-looking surface of the Moon, especially around the maria, one reader wrote a letter to the editor noting the similarity between these Moon surfaces and the surface of pure processed iron. He was an iron expert and he noted: "On examining scrap pile in an iron foundry, it is not uncommon to come across some pieces of iron exhibiting some of the structures seen on the Moon." An interesting observation, and it is striking that astronaut James Irwin of Apollo 15 described these same strangely dark areas in similar metallic terms.

Scientists took note of the great amount of metals on the lunar surface too. In an article appearing in *Science News* (January 1, 1972) entitled "Migrating Metals On the Moon . . ." it is stated: "The movement of volatilized metals over the surface of the moon could be confusing interpretation of the ages of lunar samples."

Surface metals could be responsible for yet another strange lunar phenomenon. In fact, the Moon has been tipping man off for centuries, yet very few scientists have discerned this. As we have noted in Chapter 2, strange lights have been seen on the Moon over the past several centuries. As we indicated earlier, certainly not all of these could be in any way interpreted to be UFOs. Before man went to the Moon, the speculation of some scientists trying to explain these peculiar glows was that they might be generated on the Moon's surface in some natural way. A few thought they might be caused by the peculiar surface of the lunar crust, which perhaps might possibly be loaded with metals.

There was no question that some of these glows and lights were immense. In fact, science writer William Corliss, formerly with NASA says: "The Moon . . . unquestionably luminesces over its entire sunlit face and reddish glows over areas as large as 50,000 square miles have been reported. These displays are most easily attributed to solar-induced luminescence. The highly localized, ruby red spots seen by Herschel, Kozyrev, Greenacre and many others may be either luminescence, volcanism, or some phenomenon we do not yet recognize. Only time and the landing of astronaut-geologists will tell." (*Mysteries of the Universe*.)

These glows have been verified by our astronauts. For instance, Apollo 11 astronauts did see strange glows in the crater Aristarchus, one of the areas where astronomers over the many decades have reported seeing strange luminescing on the Moon. Some described these glows as "a peculiar fluorescence on its wall."

Interestingly, this "luminous glow" seen by the astronauts was verified by several astronomic observers on Earth. The Institute for Space Research in Bochum, Germany, observed the very same glow that our Apollo 11 astronauts reported. And accounts of similar observations of the very same glows were made in Brazil, Ireland, and other far-flung areas of our planet.

Dr. H. P. Wilkins ventured this explanation for these strange glows:

"The only rational explanation would seem to be that in certain regions, the moon's surface consists of materials which emits electrons under the influence of light or electronic impact, or *considerable deposits of metals, for example, iron, which act as deflectors of free electrons*. The fluorescent effects may be due to electronic bombardment." (*Our Moon*. Emphasis added.)

Here is evidence of large amounts of metal existing in these very areas of the Moon where we later discovered them. What the great Wilkins suspected has been verified. The only question that remains is, how did it get there—not just the iron which has been found in pure form, but the other rare refractory metals that require such intense heat to melt, such as titanium.

ARE ALL THE MARIA METAL-RICH?

Soviet scientists claim that from their investigation, done mostly through remote-controlled retrieved samples, they conclude that the composition of the maria is essentially the same, although some undoubtedly have different proportions of titanium and other metallic elements.

Our own research tends to verify this conclusion. NASA document *Apollo 17: Preliminary Science Report* states clearly: "All mare basalts have been found to be unusually rich in iron and sometimes rich in titanium."

We got an inkling of this in other ways besides direct analysis of lunar material. For instance, when the Apollo 13 rocket stage (Saturn 5's third stage) crashed into the Moon (it hit with a speed of 9300 mph), the Moon reverberated, sending shock waves racing across the sphere at great speeds over vast distances. The Moon reverberated like a huge bell for over 3 hours 20 minutes.

These prolonged vibrations reinforced the view that the lunar maria are "filled with material that responds very differently to shock waves than does the material forming the Earth." (*New York Times*, April 16, 1970.) What kind of material could cause this?

We do know that material which contains heavy amounts of metal would convey tremors in such a manner. But couldn't this be a natural characteristic of lunar material? Perhaps the material out of which the Moon is made just naturally reacts this way.

On the contrary—studies show that it is just the opposite. For another complicating fact is that some lunar rocks react in just the opposite fashion. According to the highly respected publication *Science* (June 26, 1970), a study shows that the velocity of sound waves through many lunar rocks is "perplexingly low." It is, in fact, approximately just one third that of Earth rocks.

Edward Schreiber of Queens College and Orson L. Anderson of the Lamont-Doherty Geological Observatory sought out various Earth materials with a sound velocity comparable to the lunar rocks, and found that a variety of Swiss, Norwegian, Italian, and Wisconsin cheeses fit the bill.

Facetiously they point out that while the lunar rocks are about the density of cheese, this may readily be accounted for when one considers "how better aged the lunar materials are."

Their findings, they added, suggest that perhaps "old hypotheses are best after all, [and] should not be lightly discarded."

May we add, neither should any unorthodox theories—and in light of the facts, especially the unorthodox Soviet theory that the Moon may be a huge hollow spaceship!

THE IMPOSSIBLE POSSIBLE?

There is something else discovered about these strange dark regions of the outer Moon that indicates that the hand of alien intelligence was at work here. As we have already noted, evidence indicates that the maria are covered with material that is denser than the rest of the Moon. Walter Sullivan, former science editor of the *New York Times*, points out: "Furthermore, the Sea of Tranquility is covered with material that is considerably more dense than the average density of the Moon, deduced from its gravity. This is the reverse of what one would expect. On earth the lava that flows upwards and out onto the surface is the lighter component—not the heavier fraction." (*New York Times,* November 9, 1969.)

How do scientists explain how heavier materials can flow to the top of the surface of the Moon's maria? They cannot.

In fact, to be utterly frank about it, not only is this the reversal of what scientists expected, it just is not natural—not according to the way the laws of nature work. Heavier elements *sink,* not rise to the surface. The only way it can be explained is as the Soviets claim—that the maria, the dark areas of the Moon, were in some way artificially produced.

Even a little point like this, especially taking into consideration all the other evidence, leads the researcher to believe there was the hand of alien intelligence at work in the formation of the maria, even as our Soviet spaceship theorists insist.

THE GREAT OUTER SHELL OF THIS SPACESHIP MOON

The Soviet scientists Vasin and Shcherbakov stated that they were convinced that the entire outer portions of the Moon and not just the maria formed the protective outer shell of the inner spaceship world. Is there any evidence to support this?

Definitely. For now scientists know that for some strange reason the Moon has an unexplainably thick outer shell which apparently has been baked by some unknown intense heat.

This was first divined when a miracle happened. Dr. Gary Latham, who wanted to know more about the structure of our Moon's interior and the thickness of the Moon's crust, actually proposed exploding a nuclear device on the far side of the Moon to send strong shock waves all through the lunar interior. However, because of staunch opposition by other scientists and many scientific publications, he withdrew his suggestion. As it turned out, he didn't need a nuclear bomb, for a miraculous event took place. The Moon was unexpectedly hit by a gigantic meteor.

Lunar scientists with their sensitive seismometers on the Moon were awaiting such an event, hoping against hope for this unlikely hit to occur. But most scientists were realistically pessimistic, for they calculated that a meteor of such great size bangs into the Moon only about once in a million years or so. But then, on May 13, 1972, a "whopper" slammed into Luna with an impact of about 200 tons of TNT.

"It was a miracle of the first magnitude," exclaimed Dr. Latham, NASA's chief seismologist. (*Science News,* July 1, 1972.)

This miracle enabled lunar experts like Latham to determine the thickness of the Moon's outer crust. The data indicated it is about 60 kilometers (over 30 miles) thick. This shocked them. "This is twice as thick as the average crustal outer shell of any continent on Earth," notes Latham.

How did the Moon acquire such a thick, extremely tough exterior? It was somehow baked on. This in spite of the fact that our much smaller Moon (only about one fourth the size of our own planet) supposedly never could have heated up to any great extent. This is in itself a perplexing puzzle.

How then do scientists explain, even assuming the Moon could attain such high baking temperatures, such a great difference in crustal thickness compared to Earth? Again they are at a loss to say. It is just another in the long litany of lunar enigmas.

More importantly, how do scientists explain the established fact that the crust of the Moon is made of material which has been melted and poured onto the surface? Or one could put it differently: "How does one get a 65-kilometer-thick crust that is 50 to 85 percent plagioclase without melting most of the Moon? And if melting occurred how could the Moon's interior be relatively cool today?"

Conundrums that cannot be cracked. But we have seen the Soviet artificial-created-Moon theory has the answer to this and all the other lunar mysteries.

The melted and "baked" outer shell of the Moon puzzles orthodox scientists today. Asks one NASA scientist: *"What force could have brought about significant distribution of crustal material?"*

Some claim this thorough job was done by celestial bombardment, simply because their data indicate that the Moon could never have been hot enough to have extensive volcanic action. (*The Moon, An International Journal of Lunar Studies,* 1973, Vol. 8.)

However, they add: "None of the suggested artificial sources [bombardment] for the observed criterions (sic) appears to be capable of producing what was observed. . . . The only possibility would be a natural source indigenous to the Moon. . . ."

If the Moon could not generate enough heat to have done it from below and it didn't come in the form of celestial bombardment of meteors and asteroids from above, where and what force accomplished this? Could it not be a truly artificial alien force, as all the other evidence indicates?

Another news journal tells us that "there are indications that some force or forces once 'rearranged' materials on the Moon." They point to one puzzling piece of information: a rock taken from the Sea of Tranquility by our astronauts which scientists are convinced actually came from a highland area a vast distance away!

We have seen in this chapter much tantalizing and substantial evidence that indeed the Moon was the object of a massive reconversion job—that its outer surface gives clear evidence of its having been transformed from a natural asteroid or planetoid into a huge hollowed-out, reinforced spacecraft!

EXTERNALLY THE MOON APPEARS TO BE THE CREATION OF ALIEN INTELLIGENCE!

We can easily understand how all the mysteries of the maria melt away in light of the Soviet spaceship theory. Now the existence of pure titanium and other highly refractory elements found in such staggeringly great amounts in these lunar seas becomes understandable. Even the existence of pure, rustproof iron on the Moon makes sense in light of the advanced alien technology that could undoubtedly manufacture it—a process yet unknown on Earth even with our advanced technology.

Now we know how the Moon could have been heated to 4000°F.; why the Moon's maria all seem so strangely circular and appear not to be randomly distributed across the Moon's tortured face, but all on one side and, even more puzzlingly, almost all in one quadrant. This apparently was not just happenstance but planned by unknown alien intelligence.

These clearly are not naturally formed "seas." Whatever tremendous melting process refashioned the Moon world? Vasin and Shcherbakov claim it was the direct result of artificial construction by unknown alien beings. The evidence makes it appear that the Soviet scientists are shockingly correct in their speculations. Moreover, the multiplying mysteries and enigmas uncovered seem to disappear in light of this strange theory.

For instance, now the mystery of the great multitude of craters on the Moon and their puzzling shallowness becomes clear. Dr. Eugene Shoemaker, leading NASA scientist, at the outset of this lunar detective race said: "If one looked long enough and hard enough at the craters of the Moon, one became aware that 'the moon is trying to tell us something.'"

Although Shoemaker, of course, was not referring to the spaceship theory, it is clear that there are plenty of clues and a mountain of evidence to tip us off about Spaceship Moon. Now suddenly the astounding shallowness of lunar craters becomes understandable if we accept the Spaceship Moon theory—the extremely hard and tough outer shell stops all incoming blasts from celestial missiles within two or three miles of the surface.

Of course, there are other answers to this particular mystery. Undoubtedly, other planets like Mars and Mercury have broad shallow craters, which obviously does not necessarily make them spaceships either.

There are other answers to the problems of why the Moon is such a bombarded piece of rock hammered with holes numbering in the hundreds of thousands, in fact the millions, while right nearby our planet Earth escaped relatively unscathed.

Astronomers had been baffled by this particular problem until someone finally hit on a possible solution—an answer that could solve this conundrum. Admittedly, it could have happened at a time early in its cosmic career, when Earth was still hot enough to absorb the blasts, thus leaving no telltale scars, while the smaller Moon, which had already formed a hard outer shell, had holes ripped into it all over its tortured face. So even though our Earth, which was much larger and should have gotten more than its share of celestial hits, does not show that many, it is simply because the molten surface of our hot Earth at the time caused them to be effaced.

These are other solutions to these seemingly insolvable lunar mysteries that are within the realm of orthodox science and which do not require the radical spaceship theory to resolve. However, it is intriguing and impressive that all the many mysteries uncovered since scientists have put the Moon under an intense study, dissolve in light of the spaceship solution.

There is also evidence, as we have seen, that the Moon has on its surface telltale signs of artificial construction. Interestingly, the Soviet scientists are not the only ones to detect it. In George Leonard's fascinating book *Somebody Else is on the Moon* he cites various scientists outside the Soviet Union who are convinced that the Moon may be a spaceship, among them a British physicist from Oxford and a scientist at NASA's Jet Propulsion Laboratory (Cal Tech) in California.

Furthermore, Leonard claims: "At least one qualified person has argued that the skin of the Moon may *actually be an artificial protective cover—a cover which has been exposed in some places due to a horrendous debacle which took place a long time ago.*" (Emphasis added.)

Exactly what our Soviet theorists say.

Later in that work Leonard makes this shocking observation, which tends to back up this thesis: "It [the Moon] seems to have a built structure to it—a matrix, a gridwork, a weave."

He cites the fact that an astronaut dropped a piece of equipment (the LEM) on its surface and the Moon vibrated for an hour! "The shock waves were recorded by seismographs some distance away."

Amazing! Absolutely inexplicable—except, of course, in light of the Spaceship Moon theory. No one has yet offered an explanation, except the spaceship answer.

Leonard further states: "If you hit bedrock in Peking with a hydraulic hammer, I doubt it would be felt in Pittsburgh; but the same blow on the far side of the Moon would certainly be detected on the near side."

We shall see this fact reinforced when we look at the next chapter, which proves the Moon has an inner shell of metallic rock. This is our next target on the Moon.

There is nothing so far removed from us as to be beyond our reach, or so hidden that we cannot detect it.
 —*Réné Descartes (1596–1650), quoted by astronaut John Young*

TEN

DOES THE MOON HAVE AN INNER SHELL OR "HULL" OF METAL?

• Why does the Moon when close to the Earth cause a slight deviation in the magnetic needle of a compass?

• Why is there a huge bulge on the far side of our lopsided Moon, a bulge which indicates the Moon is supported on the interior to give it great internal strength?

• Why do strange circular disks centered like bull's-eyes exist in the middle of the mysterious circular maria and how does this indicate that the Moon's interior shell is "as rigid and as strong as steel"?

• NASA documents admit that there is something that exists inside the Moon that differs greatly from the rest of the interior, that is denser than rock—what they describe as "a hard layer of something deep within the Moon." What is that strange layer?

• Why does this hard layer completely enshroud the Moon?

• Why do lunar impacts cause the Moon to ring like a huge metal sphere?

• Tremors are conveyed through the Moon through its outer layers at an unbelievable rate of 6 miles per second—thirteen times faster than a rifle bullet. This is the speed of sound through metal. Does a layer of metal or metallic rock enshroud the Moon?

- The Moon suffers from "swarms" of tremors at times—what one scientist calls the "shakes." Why does this make sense only in light of Spaceship Moon's inner shell of metal?

In their mind-boggling *Sputnik* article "Is the Moon the Creation of Alien Intelligence?" the two Soviet scientists Vasin and Shcherbakov hold that the Moon is a refashioned planetoid hollowed out and reconverted into a spaceship. This cosmic Noah's Ark, they are convinced, was reinforced on the interior with a huge, thick shield of what they call "space armour," which they calculate is about 20 miles thick.

They explain: "From our point of view, the Moon is a thin-walled sphere. Probably the sphere's shell is made up of two layers—a loosely packed outer layer to absorb the shocks of meteorite impacts, with an inner layer of 20-miles thick armour-plate."

Why such a fantastically huge shell of metal or metallic rock? They explain: "Naturally, the hull of such a spaceship must be super-tough in order to stand up to the blows of meteorites and sharp fluctuations between extreme heat and extreme cold. Probably the shell is a double-layered affair—the basis a dense armouring of about 20 miles in thickness, and outside it some kind of more loosely packed covering [a thinner layer—averaging about three miles]. In certain areas—where the lunar 'seas' and 'craters' are—the upper layer is quite thin, in some cases non-existent."

So conclude our two Soviet spaceship theorists.

HARD EVIDENCE TO SUPPORT THE HARD INNER SHELL

It seems incredible that two orthodox scientists from a highly respected science institute—one of the world's most esteemed—should postulate such a strange Moon, one that according to their model has a 20-mile-thick hull! On the surface it would seem utterly preposterous.

Are there any hard-core scientific data and evidence to support this bold supposition? Surprisingly, yes.

DOES THE MOON MOVE SENSITIVE MAGNETIC NEEDLES?

Actually, our Moon gave man a number of hints that it might have just such a huge inner metal shell even before we went there.

With the interior hull purportedly being metallic and all this exterior loaded to such a great extent with metal, the question might be asked, wouldn't a metal detector on the Moon go wild?

Intriguingly, in 1861 the *American Journal of Science* (81:98–103 and 84:381–87) did note more than a century ago that the Moon seemed to have an influence on sensitive magnetic needles, causing them to deviate slightly. Their conclusion from the data and evidence then available was startling: *"The reported effects are small but definitely beyond probable error."* (Emphasis added.)

THE MOON'S PECULIAR INTERNAL STRENGTH

There were other clues the Moon offered that some scientists did not miss. For the odd shape of the Moon itself indicated to some sharp scientists on Earth that something extremely strong did exist in her interior. Even before our space probes began to rip the veil of mystery from the Moon's face, scientists on Earth surmised that our companion world was out of whack, and in fact possessed a huge bulge. However, scientists were mystified because their information indicated that peculiar bulge was seventeen times greater than could be accounted for by the tidal pull of the planet Earth. What could have caused such a lopsided Moon?

What was even more mystifying—and this scientists could not figure out—*how could the Moon manage to support such a huge bulge?* As the great lunar expert Dr. Harold Urey pointed out: "This bulge is a non-equilibrium one and *must be supported by some curious internal characteristic of the Moon, such as great strength of the interior* or some variation in the density of the body of the Moon." (*America's Race for the Moon*. Emphasis added.)

Exactly what the Spaceship Moon possesses, say our two spaceship scientists.

Then our Moon probes began and scientists got their biggest surprise. For they learned that the Moon had a bulge all right, but not exactly where scientists of Earth supposed it would be. It was, perplexingly, on the far side—a side that to our knowledge has never faced Earth. Clearly, as one scientist put it: *"It appears that, not the earth, but something else has the attention of the moon."* (*Science News*, January 29, 1972. Emphasis added.)

Although lunar experts had to revise their thinking about the Moon's lopsidedness, they knew that it still indicated considerable internal strength on the inside of the Moon. The outstanding Moon scientist Zdenak Kopal noted in 1971 that there "must be considerable strength in the deep interior of the Moon now, and that this has been true since the Moon acquired its irregular shape."

Kopal also stressed that it is surprising that a large object such as the Moon, stressed by tides and unequal heat changes, was not able to adjust its shape the few-kilometer bulge and thus relieve or smooth out this stress. That it has not done so, says Kopal, "surely indicates considerable strength in the deep interior."

One of the big questions that is perplexing lunar researchers today is, *what peculiar internal strength lies within the body of the Moon itself to support this great bulge?* Although this problem has dumbfounded scientists who today study the Moon, we now know the solution to this peculiarity of our planetary neighbor.

We now know that the "mysterious" internal characteristic that gives the Moon its great internal strength is the metallic inner shell. At least, so insist our Soviet spaceship theorists. Admittedly this mystifying puzzle would no longer perplex science in light of the Soviet theory that the Moon has in its interior a metallic inner shell or hull.

Other clues from our exploration of the Moon seem to verify that this strange world circling us has an interior as rigid and as strong as steel. When the tracing data of lunar orbiters discovered that great massive concentrations (mascons) did exist just under the surface of the circular maria, scientists were forced to conclude, as Richard Lewis puts it so well in his provocative *The Voyages of Apollo,* that "lunar rock *structure is as strong as steel."* (Emphasis added.) Indications again of that artificial construction of a metal hull inside Spaceship Moon?

So not only does the nonspherical shape of the Moon indicate great internal strength but, as this veteran science reporter puts it, the mascons also point clearly to the fact that the crustal structure of the maria indicate the lunar interior is "as rigid as steel."

ANOTHER CLUE: THE MOON RINGS LIKE A HUGE METAL SPHERE!

If such were the case—if this strange world actually had a metallic-type spaceship hull—it could have been predicted even before man went to the Moon that when our Apollo spacecraft were sent crashing into the crust the Moon would produce vibrations and tremors utterly different from any man had ever known. If the hull were really metallic all the way around, these vibrations would be of extremely long duration and the crashes should make this huge metallic hull vibrate like a huge bell or gong. Remarkably, they did!

What kind of material would convey shock waves such great distances? A metal sphere, of course, would. Whatever kind of material lies inside the Moon, there is no question that it transmits signals "very, very effectively," says Dr. Gary Latham, chief seismologist for NASA. In fact, "so effectively that we are seeing impacts from all over the Moon," Latham claims. (*Science News,* June 12, 1971.)

And these impacts are being recorded from all over the Moon, even though there are only four seismic stations placed at widely separated locations on the near side. Surely this indicates that this material completely enshrouds the entire lunar interior—whatever it is. And what material would transmit vibrations so effectively? A metallic hollow sphere could and would!

No wonder the Moon rang like a huge bell upon manmade impacts. Reconsider the data and evidence:

Apollo 12: Moon rings for nearly an hour.

Apollo 13 (third stage of rocket hit): Moon vibrates for over three hours!

Apollo 14 (lunar module hit): Moon reverberates for over 3 hours!

And on and on—until the Moon actually reverberated upon impact for more than 4 hours! *

HOW DEEP THE METAL SHELL?

The Soviet spaceship speculators concluded from their evidence that this metallic hull lay on the average about 20 miles under the Moon's hide of rock and dirt.

Consider how American evidence bears this out. When the immense Saturn rocket of the ill-fated Apollo 13 mission was propelled out of Earth orbit and sent crashing into the Moon, it produced a tremendous seismic response. The tremors traveled interiorly to a depth of 22–25 miles, according to a NASA document. (*Apollo 14: Science at Fra Mauro.*)

Scientists at NASA were naturally perplexed, but such results again are exactly what should be expected if the Soviet spaceship theory of the Moon is correct. And how correct they seem to be!—even to the amazing scientific "guess" about its inner depth.

All the other Apollo spacecraft crashes into the crust of the Moon produced similar measurable tremors through its outer covering of rock and dirt, yielding this unbelievable secret of its hidden interior. The impact, for instance, of the lunar lander *Falcon* of the Apollo 15 mission indicates, as notes the NASA report *Apollo 15,* that there is a "hard layer of something deep within the Moon." This would seem to be that inner shell of the hull of our Spaceship Moon which the Soviet scientists speculate exists there.

Scientists admit that the puzzling hours-long vibrations indicate that whatever this extremely hard layer is, it *transmits sounds extremely well.* A metallic sphere transmits

---

* "Signals from the larger man-made impacts continued for over 4 hours," *Science* magazine asserts. All signals show gradual increase and decrease in intensity. (See *Science,* November 12, 1971.)

sounds extremely well. However, NASA scientists tell us that they think it is just extremely hard rock. But our Earth has a mantle of extremely hard rock under tremendous pressure, and it does not ring or vibrate anything the way the Moon does, although admittedly the Moon apparently does not have a soft, hot interior like Earth's to dampen the length of these reverberations.

Nevertheless, not only do lunar impacts produce tremendously long reverberations but the tremors travel great distances. The signals of the Apollo 15 impact traveled over 700 miles to the Apollo 12 seismometer on the Sea of Storms and to the Apollo seismometer at the Fra Mauro highlands!

We have seen that after Apollo 12 astronauts Conrad and Bean blasted off from the Moon in their lunar module and docked with the command ship, the Apollo 12 ascent stage of that module was sent smashing into the Moon's surface. Its impact set off reverberating seismic waves in the lunar interior unlike any ever recorded on our planet. Lunar scientists reported that the seismic waves lasted over 55 minutes, the signals building to a peak in about 7 minutes and then peculiarly tapering off ever so slowly, finally dying out about an hour later. Scientists not only admit that this phenomenon indicates that the Moon's interior is an exceptionally effective conductor of vibrations but also confess that they are perplexed by the long delay between the time the signals began and the time they reached full strength, finally decaying to silence unbelievably great lengths of time later—in most instances several hours. The unusual length of these reverberations absolutely bewilders the best lunar seismic experts.

A METALLIC SHELL COMPLETELY ENSHROUDS THE MOON

Again the Moon is behaving like a hollow metal sphere. And there is evidence that the strange material that lies inside the Moon—that "hard layer of something deep within the Moon" that NASA admits is there—appears to form a complete shell that enshrouds the Moon. For the shock waves of impacts transmitted seismic responses all the way around the Moon! Similarly, as shocking as this appears to scientists, impacts actually have been received from the far side, conveyed to seismometers on the near side.

This, of course, again raises this critical question: What kind of material lies inside the Moon, enshrouding its interior, which conducts sound so efficiently and effectively that tremors are conveyed completely around the Moon? Again, that metallic layer which the Soviet scientists theorize lies inside the Moon!

THE SPEED OF SOUND THROUGH METAL

Furthermore, the vibrations indicate that the Moon has this strangely different material in an inner layer. As NASA scientific report *Apollo 16: On the Moon with Apollo 16* (1972) says, "A very recent study of the results of previous spacecraft impacts has revealed the existence of a lunar crust that may be roughly 40 miles thick. It is now believed by some of us that the Moon may be shrouded with *material that differs greatly from the material in the interior of the Moon.*" (Emphasis added.)

We have seen that the tremendous distances that the tremors are conveyed through and around the Moon indicate that it is metal. The strange prolonged tremors that perplex scientists by their duration indicate that this is a shroud of metal. But there is something even more intriguing and telling that science has learned about the inside of our Moon that is the crucial, final piece of evidence that proves it is metal. This comes from the mystifying speed with which these tremors travel through that hard inner layer that transmits vibrations so well.

As Wernher Von Braun, leading space expert, pointed out: "The velocity [of seismic waves] seems to gradually increase down to a depth of about 15 miles—then there is a sharp increase. This abrupt increase can only be accounted for by *a change to a denser material. At a depth of 40 miles, the velocity is estimated to be about six miles per second . . . No rocks examined thus far would under the actual pressures expected at a lunar depth of only 40 miles transmit seismic impulses at speeds as high as six miles per second." (Popular Science,* January 1972. Emphasis added.)

It is important to note that vibrations do travel much faster through metal, and of course through metallic rock, than through just plain rock! This data would indicate that the Moon has a layer of metal or metallic rock inside it. For we know the speed of sound through different materials. Any good physics manual can tell you this. What material would convey sounds at six miles per second—that is, thirteen times faster than a rifle bullet? Curiously, this is the same incredible speed at which the Apollo Moon-bound rockets blasted off from the Earth, bearing men to this strange, alien world. (*NASA's Apollo Expeditions to the Moon,* 1975.)

We learned on our trips to the Moon that the filled-in dark regions of the Moon are rich in metals rare on Earth —titanium, molybdenum, beryllium and the like. A check of the speed of sound through these metals so common in the outer shell of the Moon convinces us that the inner shell is composed of the same metals!

THE SPEED OF SOUND TRAVELING THROUGH LUNAR METALS*

Iron	5,960 meters/sec.	or about	18,000 feet/sec.
Nickel	6,040 " "	"	19,000 " "
Titanium	6,070 " "	"	20,000 " "
Molybdenum	6,250 " "	"	21,000 " "
Beryllium	12,890 " "	"	42,000 " "

A quick glance here reveals that an intermixture of these metals that probably exist in that inner layer would produce an alloy of metal that tremors would travel through at about 6 miles per second—in other words, about 30,000–35,000 feet per second!

We must also remember that the above chart gives the speeds of sound through ordinary metal *on Earth.* Of course, sound waves would travel even faster at a lunar depth of 20–40 miles.

* From *Handbook of Chemistry and Physics* (54th Edition), published by the Chemical Rubber Company (1973–74).

Surely this seems to be the clinching argument—the final proof that the interior of the Moon has a metallic or metallic-rock layer, that layer of our Spaceship Moon which our two Soviet spaceship scientists claim exists!

A NASA MOON MODEL IS A HOLLOW SPHERE—A SPACESHIP!

Scientists at NASA have produced various models of the Moon based on the data and information received. One such model is particularly intriguing: "Based on information scientists produced lunar models that would fit the strange phenomenon! Some such models would have made for a rather bizarre Moon, such as a hollow titanium ball. . . ." (*Science News,* November 29, 1974.)

Amazing! This is exactly what the Soviet artificial-Moon theorists claim. In commenting on this "bizarre" model Captain Lee Scherer, director of the Apollo Lunar Exploration Office, says such a model is "highly unlikely." But as we have seen, this model might be more correct than our NASA officials are willing to admit. Certainly more astounding than any scientist or human being could possibly imagine!

Yet it fits—perfectly. A huge rock-covered metal sphere? No wonder our Moon rings like a huge bell!

SCIENTISTS THEMSELVES SPECULATE THAT LUNAR METALLIC SHELL EXISTS!

Another puzzle that has scientists perplexed is why the outer portions of the Moon seem to be devoid of iron in general, while Earth is so rich in that mineral. Not that the Moon does not contain iron. Actually, as we have seen, those vast lunar plains are rich in iron as well as other metallic elements. One scientific report concludes: "From the returns of Apollo 11 and 12, geologists knew that the maria were rich in iron—one thing that makes them appear darker than surrounding areas."

But apparently there is more iron inside the Moon than scientists first thought. For now scientists are seriously speculating that a thick metallic layer lies just under the crustal surface of rock and dirt.

Remarkably, the speculation fit all the facts, for we know:

• This is the area just underneath the surface which scientists have learned is rich in metallic elements like iron. It is the area or band which, when the Moon is hit with heavy impact, sends shock waves racing all the way around the Moon. This sounds remarkably like the metallic hull of the Soviet spaceship!

• Interestingly, this layer of metal makes understandable our space agency's own conclusion that "the material that the moon may be shrouded with . . . differs from the material of the rest of the Moon." In other words, differs from the rock crust.

• This different material that enshrouds the Moon really sounds like the spaceship hull that our two Soviet scientists speculate exists just under the Moon's outer crust of rock and dirt—a material that not only transmits sounds extremely well, but which definitely gives the Moon considerable internal strength. And we do know that *something* exists inside the Moon that gives it considerable internal strength to allow for the existence of the nonequilibrium figure of the Moon—the lopsided bulge of that orb. This spaceship hull or shell could be that "something."

• The speed at which tremors travel through this internal layer is the speed at which sounds travels through metals. In conclusion, every indication exists that here seems to lie the metallic hull of our Spaceship Moon.

TECHNICAL DATA SUPPORT THE METAL-HULL THEORY

Furthermore, various other lunar studies reveal that just such a layer of metal does exist beneath the outer rock crust of our lunar world. In a technical study reported by the NASA-sponsored Fourth Lunar Conference (*Proceedings of the Fourth Lunar Conference,* Pergamon Press, 1973, Vol. 3) Dr. Curtis Parkin, Department of Physics, University of Santa Clara, and Palmer Dyal and William Daily of NASA's Ames Research Center, using Apollo 12 and Apollo 15 lunar-surface magnetometer data with simultaneous lunar orbiting Explorer 35 data, plotted hysteresis curves for the entire Moon. In their technical paper, "Iron Abundance in the Moon from Magnetometer Measurements," these lunar experts declare:

"This result implies that the moon is not composed entirely of paramagnetic material [rock] but that ferromagnetic material such as free iron exists in sufficient amounts to dominate the bulk lunar susceptibility."

The study's conclusion is staggering in its implications: "We have found no reasonable paramagnetic mineral or combination of minerals with the correct lunar density that would have permeability high enough to be consistent with our measured values. From this we infer that the moon must contain some material in the ferromagnetic state, such as *free metallic iron,* in order to account for the measured global permeability." (Emphasis added.)

Other studies from different data have come to a similar conclusion: that metallic material exists in a large amount in a layer just under the outer rock surface of our Moon. The highly regarded lunar journal *The Moon, an International Journal of Lunar Studies* points to three such studies: Sonnett's (1971), Urey's (1971), and Murphy's (1971). They all point to a metallic layer inside the Moon. Sonnett postulated a three-layered structure for the Moon similar to the Soviets', a special feature of which was "a thin conductivity layer . . . which has been suggested to be Fe [iron] metal" or a similar alloy.

THE OUTER MOON IS A STRANGE HUGE RUBBLE PILE

Another perplexing mystery of the Moon is the fact that the outer parts seem to be a huge rubble pile 20 miles deep. Dr. J. A. Wood of the Smithsonian Astrophysical Observatory noted at the Apollo 14 pre-mission science briefing:

"It's from this layer of broken-up rubbish that the astronauts make their collection." (J. A. Wood et al., "Lunar Highlands Anorthosite and Its Implications," *Proceedings of Apollo 11 Lunar Conference,* Pergamon Press, 1970.)

Dr. Gary Latham, chief seismological researcher, points out: "The evidence seems to indicate a broken up, rubble structure in the outer 10 or 20 miles." What could have created this? "We are not able to say how this structure came into being."

Another mystery that can be solved in light of the Spaceship Moon with its inner hull of metal or metallic rock. For if the Moon does have this hard inner shell, then the outer rubble pile makes sense. Most scientists like Wood agree that the outer layer of broken hard rock under the topsoil was probably created by the unrelenting pounding of meteors, asteroids, and comets striking the lunar surface over eons. But why should it break up? Because after eons and eons of the constant pounding Spaceship Moon obviously received as it traveled through the cosmos, gradually but persistently the outer layers of the Moon would be broken up as great meteors and asteroids (undoubtedly including a comet or two) crashed into its crust. Like candy coating covering a harder inner shell, the outer coating of rock gradually began to crack and break up, leaving what we see today—a huge rubble-pile structure in its outer crust which is 10–20 miles deep.

Spaceship Moon solves yet another lunar conundrum.

THE MYSTERY OF SWARMS: "THE MOON HAS THE SHAKES"

Before man went to the Moon most scientists were convinced it was a dead world. Until recently, as far as man could tell it was a very quiet world. But after man landed and set up various seismological stations he soon discovered that it was a very active world indeed. Swarms of tremors or activities have been discovered to be taking place inside this mystery world. Not only moonquakes or activities deeper than any detected on Earth, focused 500–1000 miles beneath the lunar surface, but something else that is very strange was detected more toward the outer surface.

Peculiar swarms of seismological activities are being picked up by our sensitive seismometers. Swarms of what scientists call "mini-quakes" for want of a better term have been detected. These reports of internal activities or tremors are over and above the larger, presumably tiny moonquakes that are reported. Chief seismologist Gary Latham of Columbia University Geological Observatory, Palisades, New York, along with his expert staff has been analyzing these strange unconventional seismic signals, which seem to come, as he describes them, in "swarms."

Latham notes: "During the period of swarm activity, events occur as frequently as once every two hours over intervals lasting several days. The source of swarms is unknown at present." (*The Moon, An International Journal of Lunar Studies,* 1972. Also see *Science News,* November 29, 1971.) Another Moon mystery.

Instruments have recorded transmissions of high-frequency waves lasting from one to nine minutes. Scientists are befuddled, although a few conjecture that they might be landslides in the Moon's outer surface areas. They continue on and on, so this is an unlikely explanation.

Literally thousands of these tiny tremors have been reported, all of them believed to be of natural origin. Latham noticed one distinct pattern for these shakes and quakes: Nearly all of them occur each month when the Moon comes closest to the Earth. He naturally infers that the increased tug on the Moon by the Earth's gravity produces tidal stresses on the Moon, causing some shifting and shaking of its outer crust or shell. And the Earth's tug on the Moon is considerable, much greater than the Moon's pull on Earth.

Interestingly, the moonquakes occur most frequently along the lunar rills, those narrow but mysterious deep canyons that meander for hundreds of miles across the lunar surface, which some scientists (as we noted in *Our Mysterious Spaceship Moon*) are convinced "cannot exist."

The existence of moonquakes or internal seismic activity does not mean, Latham points out, that the Moon's interior is necessarily hot. They do not seem to be due to volcanic activity but rather to the shifting of the crust of the Moon.

These weak tremors are for the most part less than magnitude 2 on the Richter scale. They seldom reach depths of more than half a mile.

Incidentally, this low level of seismic activity is another indication that the Moon's outer shell is rigid and stable, far more so than our Earth's. So conclude lunar experts like Latham.

These myriad tremors have led one scientist to observe that it almost seems as though the Moon has "the shakes." The cause is intriguingly mystifying. Could it be, however, that some of these swarms of tremors are caused by the cracked, broken crustal layers of rock which when out-of-balance are adjusting themselves over the hard continuous inner shell of Spaceship Moon? We do know, as we have seen, that just such a shroud or band of extremely hard material that conveys tremors extremely well and appears to be metallic does exist there. Could it be that as the Earth's pull causes the outer crust of the Moon to shift constantly back and forth over this hard inner shell seismic evidence says the Moon has the shakes? The Moon's huge outer rubble pile's continually shifting seems to be the cause. And the solution seems to be the hull of our Spaceship Moon.

ALL SEISMIC AND LUNAR EVIDENCE = HOLLOW METAL-HULLED SPACESHIP!

As we have seen, the entire scientific picture painted by seismic signals as learned from the Apollo lunar program indicates that our Moon is an artificial world inside—a spaceship, as the two Soviet scientists maintain.

We have also cited various studies done by a whole host of scientific experts to back this frightening conclusion.

Finally, we have noted how such a metal-hulled spaceship dissolves some of the sticky and perplexing problems that have mystified lunar scientists ever since man began to study our neighboring mystery world.

No wonder some NASA scientists have come up with a model of the Moon that is a rock-encrusted hollow metallic sphere. Such a Moon fits the evidence. And such a Moon is exactly what our Soviet scientists claim our satellite to be—nothing else than a huge hollowed out, alien-created spaceship!

The moon is teaching us extraordinary things . . .
—*Dr. Gerald Wasserburg, Cal Tech, NASA scientist*

ELEVEN
ARTIFICIAL CONSTRUCTION INSIDE THE MOON?

• What would cause a *"Straight Wall"* Formation on the Moon's exterior—an unusual mountain or cliff-like "construction" that extends for 60 miles?

• What are the strange circular disks centered like a bull's-eye in the middle of the Moon's circular seas? Why do scientists say they should not be there—in fact cannot exist?

• What evidence is there that gigantic girderlike blocks of metal about 1000 kilometers long exist inside the Moon?

• How do scientists explain identical seismographic signal trackings that keep coming from the Moon's interior? Why is this seemingly impossible? How does it indicate artificial construction in the Moon's interior?

Astronomers in the past actually claimed to have detected strange "constructions" on the outer surface of our satellite. For centuries they stared and strained their eyes through Earth telescopes, sometimes perceiving strange sights on the lunar surface. The German astronomer F. Gruithuisen insisted that he saw what he referred to as "ramparts" on the Moon—artificial fortresses built apparently by some kind of Moon beings. He even claimed that he saw the roofs of their houses and could make out their roads.

Although Gruithuisen undoubtedly had an overactive imagination, he was considered a competent astronomer and made several notable contributions to our general knowledge and understanding of the Moon. Similarly, there is no doubt that most of the so-called artificial constructions detected on our Moon's surface were actually mistaken interpretations of natural features. However, as we have already seen (and shall see again in a future chapter), apparently a number are almost certainly some kind of artificial construction.

Furthermore, as the late H. P. Wilkins observed, it is hard to believe that these competent observers were completely mistaken, that what they saw was purely a product of their imagination. In some cases they most certainly saw something. What that "something" was we can today only guess.

THE STRANGE STRAIGHT WALL ON THE MOON

One of the natural features of the Moon that perplexed astronomers in the past even as it causes wonder among moongazers today is the strange straight structure known as the "Straight Wall." One astronomy guide says that it is a "feature so artificial in appearance" that in the late nineteenth century it was frequently referred to as "The Railway." Some astronomers observing this strange structure, which seemingly rises from the bowels of the Moon, going steadily straight outward (although not perfectly straight for over 60 miles), claimed that it could be nothing else but a "constructed wall" made without a doubt by intelligent beings. This was a common opinion in the seventeenth and eighteenth centuries.

Today, of course, astronomers unanimously disagree with this interpretation; the "Straight Wall" appears to be merely a huge fault line. Still, all agree that it is most unusual—it looks almost as if something underneath were pushing up through the crust of the Moon.

Exactly! say Vasin and Shcherbakov. For our Soviet spaceship scientists speculate that it is not a surface construction, although it is not completely a natural feature either. It is strangely straight, they hold, because this most splendid feature of the lunarscape—a straight "wall" nearly 500 yards wide and over 60 miles long—"formed as a result of one of the armour plates bending under the impact of celestial torpedoes and raising one of its straight even edges."

It is in effect a section of the spaceship hull or shell that has been ruptured or otherwise damaged that creates this unusual feature. It is interesting to note that today the outline of a huge, 100-mile-wide crater caused by a meteor which impacted here eons ago right over this very area can be seen. Could this have unhinged the spaceship hull, as the Soviet scientists speculate?

Frankly, this bizarre proposal appears to be sheer guesswork; yet, strangely enough, on the back side almost directly opposite the "straight wall" is a huge crack, 150 miles long and 5 miles wide in some places. Is it too farfetched to speculate that if the Moon does have the inner metallic hull of a spaceship—and there is every indication that it does—this vast opening is somehow related in its formation to the Straight Wall? Could it be that when the bow plating of the hull was ruptured or damaged, thus pushing up the huge fault line, as our Soviet spaceship theorists speculate, at the same time it pulled open this great crack formation on the far side of the Moon?

The huge circular seas of lava, which are apparently loaded with metals, would have required gargantuan equipment to construct. If this is correct, the Soviet scientists hold, they could be ponderous enough to cause gravitational anomalies that could be detected by man.

Amazingly, as our early unmanned spacecraft were orbiting the Moon, they detected such gravitational anomalies over the maria! In 1968 tracking data for the lunar orbiters first indicated that massive concentrations (mascons) existed under the surface of these circular maria. In fact, the gravitational anomalies were so pronounced that the spacecraft actually dipped slightly and accelerated when passing over the circular lunar plains.

Irving Michelson notes: "When the spacecraft passed close to these buried masses, the resultant lunar gravitational attraction it experienced was abruptly and drastically modified, thereby changing the subsequent path in much the same manner as if engine thrust had been applied." Eugene Rabinovich, ed., *Man on the Moon*, Basic Books 1969.)

Scientists calculated that the cause was "enormous concentrations" of dense, heavy matter which, as one scientist graphically phrased it, was "centered like a bull's-eye under the circular maria."

Scientists at first speculated that these mascons were just intense concentrations of lava that filled the deep basins. Opponents of this theory claimed that there was just not enough lava there to produce the effect the mascons were giving.

Other lunar experts thought that the mascons might be the remains of huge meteors or asteroids that hit here when the maria were created. However, this offers even more difficulties, not the least of which is, as we have seen, the solid evidence that the circular dark plains of the Moon were not caused by celestial bombardment but by massive upwelling from some internal source.

We can only speculate and guess. It may almost seem as if in this "guesswork" we were weaving science fiction here. But as we have seen, the guesses at least to some extent rest on solid scientific fact.

WHAT LIES INSIDE SPACESHIP MOON?

Although astronomers in the past have had a field day observing strange things on the surface of our Moon, no one has speculated that artificial constructions of all kinds have existed *inside* the Moon. That is, until Vasin and Shcherbakov came along with their Spaceship Moon theory.

These two Soviet Academy of Sciences researchers speculate that inside the Moon all kinds of constructions, including machinery and engines, must exist yet to this day. As they note: "In other words, everything necessary to enable this 'Caravelle of the Universe' to serve as a kind of Noah's Ark of intelligence, perhaps even as the home of a whole civilization envisaging a prolonged (thousands of millions of years) existence and long wanderings through space (thousands of millions of miles)."

If they are right in this bold idea, and the Moon is actually a spaceship, naturally the machines and artificial constructions, whatever they might be, undoubtedly still exist inside our Spaceship Moon.

Amazingly, from seismic study and the data of very

sensitive instrumental detection comes evidence that some kind of artificial construction does exist inside Luna.

MACHINES FOR MAKING THE MARIA?

The authors of the Spaceship Moon theory, who believe that tremendous amounts of metallic rock were poured forth onto the surface in the creation of a hollowed-out moonship world, also are convinced that "the stocks of materials and machinery for doing this are no doubt still where they were." They also claim that scientists have detected them. It should be noted, however, that they are only speculating that this is what they are. Here is another possibility, expressed by Irving Michelson in *Man on the Moon:* "Cassini's assertion of the perfect uniformity of the moon's rotational motion might also be explained by the mascons, since their effect would be to contribute to a lunar structure particularly conducive to the extremely stable motion that the observed regularity represents. Newtonian mechanics and gravitation certainly also stand firm in that new picture, and the formerly inexplicable high values of the latest inertia moment difference can likewise be fully accounted for by the presence of the mascons."

Note also that the mascons, whatever they may be, are huge circular disks centered like a bull's-eye smack in the center of each circular maria. Scientists have pointed out that they could not be the "remains" of celestial meteors, for those would have vaporized upon impact, since they hit with enough speed to cause fierce explosions equal to the force of any atomic bomb.

Scientists are perplexed and befuddled by the mascons in more ways than one. In fact, ever since their discovery in the late sixties the mascons have proved to be a major problem. As one scientist has put it: "No one seems to know quite what to do with them."

Actually, scientists did not expect to find such gravitational anomalies under the billiard-table-like lunar plains. Our early Apollo expeditions carried out a series of experiments verifying their existence. Walter Brown of the Jet Propulsion Lab at Pasadena confessed: "We expected on the marias signals would reveal smooth subsurface structures and the opposite would be true for the highlands . . . but the sounder signals were well-behaved over the Highlands and bounced around over the maria, implying subsurface features . . . disc-like objects."

What they are is a major Moon mystery. It now appears that the mascons are broad disk-shaped objects that could be possibly some kind of artificial construction. For huge circular disks are not likely to be beneath each huge maria, centered like bull's-eyes in the middle of each, by coincidence or accident.

But there is yet another mystery with the mascons beyond just their unsuspected, unlikely, and hard-to-account-for existence—a mystery which may furnish us with still another clue proving them to be not completely natural. The mascons are situated in seas of metallic lava, and it is perplexing that they have not sunk away from the surface. Any way scientists look at the maria they appear to have been formed by heat. Why didn't they sink to the bottom of these molten lava seas before the lava hardened?

Dr. Michael Yates, a leading NASA scientist, summarizes this problem: "There is a serious problem with the mascons, with any theory that says the moon is 'hot.' For there is no reason why the structures would not sink into a hot moon, unless that inner structure cooled soon . . . or unless there is some insulating region between the outer and inner areas." (*Science News,* April 3, 1971.)

Of course, if they are some kind of artificial construction, as Shcherbakov and Vasin suggest, then these problems disappear. In the orthodox view the mascons should not be there!

Then what in the name of Apollo are they? Perhaps the answer to this puzzle might be found in terms of intelligent alien formation of the mascons, as the Soviet spaceship scientists insist.

THOUSAND-KILOMETER-LONG GIRDERLIKE BLOCKS OF IRON IN THE MOON?

If there is that much artificial construction and "all manner of machinery," as the Soviet scientists suggest, wouldn't our sensitive seismometers on the Moon have detected them?

Surprisingly, they have! From the data of our sensitive seismometers comes a report that indicates that artificial construction of some kind does definitely exist inside Luna. A science report from the office of Dr. Gary Latham, NASA's chief seismologist, reveals that two "belts of activity deep inside the Moon have been detected through our sensitive seismometers."

The report claims that "the belts are at least 1000 kilometers long and 1000 kilometers deep and do not appear to intersect." (*Science News,* April 7, 1973.) It observes: "The most exciting and puzzling aspect of the active zones (or belts) is their distribution."

Dr. Latham frankly admits he is puzzled by the entire discovery. "I'm mystified. We can't explain it yet."

Latham rules out the possibility that they are great fracture systems, since the quakes in a given belt "do not show a systematic correlation with lunar tides." "Another possibility is that they are *composed of material such as embedded blocks of iron* that would cause them to have different elastic properties from the rest of the Moon." (*Science News,* April 7, 1973. Emphasis added.)

Embedded blocks of iron found in natural form inside the Moon? It is not explained how this is possible, although the report implies that since the belts are 1000 kilometers long, the "embedded blocks of iron" are also about that long!

Obviously, there is no way iron blocks like that could exist *naturally* inside the Moon. The thousand-kilometer-long blocks of metal simply do not exist naturally. In light of the Spaceship Moon theory, huge girderlike blocks of metal become not only understandable but something that should be expected.

THE MYSTERY OF MOON MYSTERIES— IDENTICAL SIGNAL TRACKINGS

Unbelievably, there is even further evidence of artificial construction inside our Moon. This evidence again comes from our seismometers on the Moon. For our scientists have found an even more perplexing puzzle—this one absolutely mind-boggling—and that is the preciseness with which moonquakes are triggered at not only the same time but, mystifyingly, in the same way. The seismographic readouts indicate they are always identical!

For perhaps the most striking phenomenon of all regarding our strange Moon is that the seismic recordings each month conform to nearly the same identical pattern. The sequence of events, as Earth's gravity tugs at the Moon, seems to be the same each month, rendering the same "seismic signature." Frankly, what is happening appears to be impossible!

Lunar seismic disturbances are assumed to be generated by stresses building up and sliding or other movement of rock faces. In the case of the Moon, lunar expert Gary Latham explains, probably the tidal pull builds up until the "friction can no longer hold these surfaces together and they just pop; they slide. . . ." But of course they would not slide each and every time in the same way at the same time. And yet they appear to be doing just that! The happenstance shifting of differing rock layers performing in the same way appears to be impossible. Dr. Latham admits this phenomenon defies explanation.

Moonquakes occur at monthly intervals like clockwork. When the Moon is closest (perigee) the first popping noises come. Actually, the very first occur five days before the Moon reaches perigee in its orbit, and then again another event indicates something stirring inside the Moon three days before perigee. The amazing thing is the clockwork precision with which this all happens. Scientists find it absolutely astonishing. "You can nearly set your watch by it," confesses Latham. (NASA Science Briefing, Houston, May 26, 1971.)

Even stranger than the identical timing is the phenomenon of the identical signal trackings scientists are receiving from our lunar seismometers!

It is understandable that the gravitational pull of the Moon can cause the tides and even uplift the solid Earth, causing it to bulge a few inches at high tide. Earth exerts a much greater pull on the Moon—in fact, twenty times as much. This is what makes the result so baffling. Seismic signals being received at different seismological stations on the Moon are identical for each respective station.

A *New York Times* report spells out this perplexing puzzle: "A remarkable recently discovered feature of these events is that each one transmits a complex but identical sequence of signals through the Moon. If one lays the records of tremors recorded at one station for a number of these events above and below one another, they are virtually the same!" The report adds: "The fact that these events are always identical is remarkable, according to seleneologists." (*New York Times*, August 4, 1972, and April 27, 1971.)

That is the understatement of the century. For how could the shifting of rock and crust be always the same? How could they produce identical seismographic recordings?

New York Times editor Walter Sullivan compares the likelihood of this happening to the possibility of the stock market always registering the same each and every time. "It is as though the ups and downs of the stock market repeated themselves precisely for each period of fluctuation." Impossible? How could each seismographic station on the Moon come up with virtually identical signal sequences for that station, time after time, month after month? The records for Apollo 12 site are thus all the same; those at Apollo 14 are the same.

What could cause identical seismic signal trackings from inside our Moon? Latham confesses he has no explanation. He is utterly mystified. Latham admits that earthquakes naturally produce different seismic recordings each and every time. There are *very rare* instances of identical recordings made in the Himalayan Mountains—but only rare instances, not each and every time, as they are inside the Moon.

In our opinion, it is hard to understand how this could be a natural phenomenon. However, something artificially constructed could produce the same identical seismic result, which could occur over and over. Could not this be the answer to this mystifying mystery of our Moon?

Could not, for instance, two long girderlike blocks of metal pulled by the gravity of Earth move the same way each and every time, thus producing identical signal trackings for each respective seismometer on the Moon?

Absolutely. In fact, you would expect it. This is, in our opinion, striking evidence that the Moon does in fact have such artificial construction! No other explanation seems plausible.

Of all the major pieces of evidence proving the Moon is a spaceship this we are convinced is the single most important. The implication of this evidence is clear—the Moon does have artificial construction on the inside. It is a spaceship. What other answer can there be?

The more we see of the moon, the more complicated we know it is.
—*Dr. Robin Brett NASA scientist*

TWELVE
THE MOON IS NO LONGER A MYSTERY

What further evidence indicates that the Moon is a spaceship?

• What evidence is there that the Moon is much older than Earth and therefore came from some other corner of the universe?

• Why is it that some scientists claim that Moon rocks have been found which were dated up to 7 and even 20 billion years of age?

• Did NASA really reveal a rock that was 5.3 billion years old—nearly a billion years older than the estimated age of Earth and solar system?

• Why did one Nobel prize-winning scientist and leading lunar expert claim that certain elements found on the Moon indicate that the Moon is much older than Earth but that he cannot explain how it got here?

• Why were rocks that were dated 4.4 and 4.6 billion years old called "the younger rocks on the Moon" by leading lunar experts?

• Why do great amounts of argon 40 found in lunar samples lead scientists to conclude that the Moon must be at least 7 billion years old—almost twice as old as our own Sun and Earth?

• Why does the lunar soil appear to be a billion years older than Moon rocks? And why is this on the surface seemingly impossible? Why does the Spaceship Moon theory dissolve this "impossibility"?

• Why did a group of scientists propose that the Moon was formed *between the stars before our Sun was born,* and later captured by Earth? How was it then drawn into a stable, circular orbit around Earth? Why is this most difficult to explain?

• How does the mystifyingly different chemical composition of the Moon indicate that it was not formed in its present orbit around our Earth but somewhere else in the universe?

• Why does the Moon appear today to scientists to have been "made inside out," as one leading NASA scientist put it? Why does this exactly fit the hollowed-out-spaceship theory?

• Why are the Moon's upper 8 miles so highly radioactive, loaded with elements that should not be there *naturally?*

• How can NASA scientists explain the huge cloud of water vapor 100 miles square that was discovered on this dry, dry world? What evidence exists that it definitely came from the Moon's interior, and why does this perfectly fit the spaceship theory?

• Clear evidence exists that the Moon was once a hot body; equally valid evidence exists that the Moon could never have been that hot naturally. Why is it that scientists cannot resolve this contradiction, and why is it that the Soviet spaceship theory does?

• Why does the Moon yield so many contradictions and paradoxes, and why is it that all are easily understandable in terms of the spaceship theory?

Long ago it was thought that standing in Luna's cold

white light too long could cause you to go mad—to become a lunatic! Undoubtedly there are people who hear me enthusiastically proclaim that the Moon may be a spaceship and think: "Wilson, you have been standing in the light of the full Moon too long! You and your Soviet theorists are lunatics!"

Perhaps when you first heard of the Russian theory you thought the same thing. But now that you have covered the latest mind-boggling Apollo findings and see how everything fits the spaceship picture so neatly, maybe you have found your mind bending a bit.

Former NASA science researcher William Corliss in his book *Mysteries of the Universe* observed before our astronauts went to the Moon: "The astronauts of Project Apollo are journeying to a world that is radically different from ours. They should bring back not only answers to our questions about lunar activity but mysteries far deeper than 'mere lights on the moon.' "

Little did Corliss realize how prophetic his words would be! Instead of giving Earth's scientists a clear picture of the origin and nature of our neighbor, the Apollo missions merely added mystery upon mystery, until now it seems that science finds the Moon a complete enigma.

In my search for the truth about our Moon—whatever it might be—I came across a fascinating article in the *New York Times Magazine* (May 1972) entitled "The Moon is More of a Mystery Than Ever," written by Earl Ubell, then science editor of CBS News in New York. He stated that our Apollo findings swirled with contradictions and mysteries. Some facts indicated that the Moon was formed cold; others that it was hot; some that the Moon does not have a magnetic field, others that it once did, and so on.

In fact, probably the most pervasive findings in all our lunar research was the discovery that the deeper scientists probed, trying to unravel the truth about the Moon's make-up and origins, the more confused they became. Equally valid data and evidence seemed to back contradictory conclusions. About the only thing for certain the scientific probings produced was just more uncertainty —more questions and more and more mysteries. In fact, instead of giving us answers to key questions and perplexing lunar problems, as one NASA expert Dr. Gerald Wasserburg put it: "The Moon is giving us answers we don't even have questions for."

Over 800 geologists, chemists, geochemists, geophysical and astrophysical experts, physicists and astronomers, including all the greats in those fields, have studied the Moon evidence, the rock and soil samples and the other information, in a valiant attempt to learn the truth about the perplexing satellite orbiting our Earth.

These giants of the intellectual world have undertaken perhaps the toughest detective job ever tackled by man. But as Columbia University's Dr. Gary Latham, one of NASA's leading lights, confessed, it nevertheless is "the most exciting business in the world." This despite the fact that its results have turned out to be mostly frustrating with conflicting, contradictory evidence and clues all over the place confusing even the most knowledgeable lunar experts.

Ubell summarized the situation in his article: "Most moonmen believe that there is a Kepler or Copernicus, or a Darwin, or an Einstein waiting in the wings to take all the contradictory data and theory and weld them into one explanation. That synthesis could have as great an impact on the intellectual and ordinary life of man as did the ideas of a sun-centered world, the theory of human evolution and the theory of relativity."

If the Soviet spaceship theory is confirmed, then indeed Ubell would be even more correct than he dreamed!

The amazing outcome, as we have seen, is that the lunar evidence indeed appears to do just that—it proves that this strange world of the Moon is a hollow and therefore artificial satellite—in short, a spaceship!

For, remarkably, the myriad mysteries of the Moon uncovered by our continuing lunar space probes appear to disappear in light of Spaceship Moon. In fact, amazingly, they are not mysteries at all—for under the artificial Moon theory they are *actually what would be expected*.

In his *Times* article Ubell lists a number of mysteries of the Moon that today perplex lunar experts. Let us review them in light of years of matured scientific detective work to see if they indeed do disappear when scrutinized through this Soviet artificial-Moon theory.

The moon did have a far older surface than the earth.
—Dr. Robert Jastrow

MYSTERY #1: THE AGE OF THE MOON—IS IT FAR OLDER THAN EARTH?

When our astronauts first went to the Moon and brought back rocks for scientists to examine and analyze, scientists never expected to find the Moon rocks far older than the Earth or even the solar system.

As Earl Ubell tells us: "Everything about the moon is old ... old ... old. Scientists certainly expected things on the Moon to have been relatively undisturbed, but they were not prepared for the pervading antiquity."

Amazingly over 99 percent of the Moon rocks brought back turned out to be older than 90 per cent of the oldest rocks found on our planet Earth. Some rocks, a few scientists insist, are far older than even our star, the Sun.

The very first rocks that astronaut Neil Armstrong picked up at random after landing on the Sea of Tranquility turned out to be more than 3.6 billion years old! This is quite remarkable in light of the fact that our scientists on Earth have searched for centuries to find rock even near that old.

The oldest rocks ever found on Earth up to recently were around 3.5 billion years old, and these were searched out of some African crannies. A few slightly older, now purportedly the oldest rocks in the world, have since been discovered in Greenland. They are dated around 3.7 billion years old—approximately the same age as the rock from the Sea of Tranquility.

But this was only the beginning of the Lunar antiquity story. One rock from man's first trip to the Moon turned out to be a baffling 4.3 billion years old. Another puzzler (Rock 13) checked out at an unbelievable 4.5 billion years of age. Finally, an Apollo 11 soil sample turned out to be 4.6 billion years old—the estimated age of the entire solar system itself. Strangely, the soil was about a billion years older than the rocks around it.

But this entire story is much more remarkable than it appears on the surface. For scientists had evidence that the maria were the younger parts of the Moon, so that the rocks found there would be among the *younger* rocks of the Moon. In other words, as science reporter Richard Lewis puts it, "rocks as old as the oldest ever found on Earth were actually the younger rocks of the Moon!" (*The Voyages of Apollo.*) Amazing. Unbelievable.

Soviet unmanned landings on the Moon have yielded similar results. The Soviet probes have returned with samples of that alien world from several of her "seas." One of the Moon's older seas—or so it appears—turned out to be the Sea of Fertility. For the data indicated that it was 4.6 billion years old—as old as the solar system itself. But scientists have a conundrum on their hands, for other equally valid evidence indicates that such a sea actually is the youngest part of our planetary companion world. How is this possible?

The age of our planet Earth is now generally agreed to be about 4.6 billion years—the age of the Sun and its planets. The best estimate of the age of the solar system derives from the age given to meteors—about 4.6 billion years old. Yet there is solid evidence that some Moon rocks and lunar soil itself indicate they're far older than meteorites. Richard Lewis summarizes the problem: "The meteorites are the Rosetta Stones of the Solar System. They are dated 4.5 billion years. Since they are composed of primitive material, they are believed to be the earliest condensates from the solar nebulae." (*The Voyages of Apollo.*)

Unquestionably, the evidence indicates that something was askew with the Moon. It appeared that the Moon just did not fit into our solar system family. But most NASA scientists stubbornly refused to believe that the Moon could be older than meteorites or Earth, let alone the solar system. They clung to their old orthodox view unwilling to allow their old theories to die in the face of overwhelming evidence. If the evidence does not fit, discard it, seemed to be their rule. Yet admittedly, early in the Apollo program, a NASA scientist stated the problem clearly: "The Moon is 4.6 billion years old, as old as the solar system itself and perhaps even older than the Earth."

Dr. Harold Urey, highly respected lunar expert does

note, however, that when we say the Earth is that age we are, of course, making an assumption. Urey points out: "We have no proof." Urey, as we shall see, was one of the scientific experts who came up with solid evidence that the Moon is much older than our Earth or our solar system. Proof which apparently has not been yet accepted—for NASA still clings stubbornly to that 4.6-billion-year limit. As Dr. Gerald Wasserburg, a jovial scientific scholar with a wit as sharp as his razorlike mind, responded when asked what hypothesis he favored concerning the Moon's approximate age:

> Some like it hot,
> Some like it cold,
> Some like it in the pot,
> 4.6 billion years old.

However, Wasserburg was so confused by the puzzles and problems of dating lunar materials brought back by our astronauts that he facetiously referred to his own scientific lab at Cal Tech as the "Lunatic Asylum" and himself and his co-workers as "inmates." Stuffy orthodox scientists thought this was in poor taste and criticized him for it. Undaunted Wasserburg ignored the criticism and put a brass plate on the door of his lab. It reads: LUNATIC ASYLUM.

Wasserburg expresses a serious point in his lighthearted banter poking fun at himself and his lunar colleagues. The Moon with all its paradoxes, contradictions and perplexing conundrums is driving scientists crazy.

Although NASA scientists claim agreement that the moon is no more than 4.6 billion years of age, still, according to Harvard's *Sky and Telescope*, a highly respected astronomy journal, the Lunar Conference of 1973 revealed a moon rock that had been dated at 5.3 billion years old. Shockingly, this is nearly a billion years older than the oldest estimate ever given to Earth and the solar system itself.

Yet if we are to believe other claims, much older Moon samples have been brought back from this strange world circling our skies. For another report claimed that based on the potassium-argon system of dating, now accepted by science as the most accurate dating system, "some of the rocks gave an unacceptable age of 7 billion years." (Editors of *Pensees, Velikovsky Reconsidered,* Doubleday, 1976.) This same journal refers to yet another study which asserts that "two Apollo XII rocks have been dated at 20 billion years of age." This would be about four times the age of our planet! To this author's knowledge, this is as old as the very oldest scientific estimate ever placed on any portion of the universe!

Admittedly, as science writer Richard Lewis points out in his definitive work *The Voyages of Apollo:* "The dating game scientists played with the lunar rocks was characterized by discrepancy, inexactness and contradictory measurements." Lewis notes that tests done to measuring the rocks' decay of uranium and thorium led some scientists to come up with older figures than those obtained by potassium-argon or rubidium-strontium decay methods. Says Lewis: "Unless there was error, this implied that the uranium-lead system data were revealing materials more primitive than the rocks they formed."

The most popular way of explaining it, NASA Moon expert Leon Silver explains, "would be to suggest that the isotope systems of uranium, thorium and lead have a memory of some prior stage of rock evolution." (*The Voyages of Apollo.*)

Other evidence tends to prove that the Moon is much older than Earth and therefore had to come from some other corner of the universe. Dr. D. Heymann, expert geologist at Rice Institute, who examined the lunar surface soil, said that scientists found "an unusually large amount of argon-40, an isotope of the noble gas argon." Walter Sullivan, former science editor of the *New York Times,* said that the surprising amount of this argon gas in Moon samples has shocking implications for man's Moon. "The Moon would have to be 7 billion years old," Dr. Heymann figured, to account for "an accumulation of argon-40. . . ." The implication here, of course, is that our Moon has not always been circling our planet. (*New York Times,* January 7, 1970.)

Dr. Harold Urey came up with similar findings based

on different evidence furnished by returned lunar samples. Urey claims that certain elements have been discovered on the Moon which indicate definitely that the Moon is much older than our Earth. Writing in the technical science journal *Chemistry*, Urey claims that "moon rock has been shown to contain xenon isotopes from fission of plutonium-244 which are not found on earth. . . ." This "indicate[s] that the moon is much older," concludes Dr. Urey. (*Chemistry*, February 1974.)

If this is true, it would of course mean that the Moon came from somewhere else and was captured in Earth's gravitational field sometime in Earth's past. Urey agrees, noting: "But despite the evidence, most scientists feel the capture theory is unlikely." (*Chemistry*, February 1974.)

With all this weighty evidence on their hands, it is not surprising that rather early in our Apollo lunar program some scientists formulated a theory that would fit the Soviet spaceship hypothesis. According to Walter Sullivan in his article ("Some Startling Findings on Those Moon Rocks"): "Another proposal has been made that the *Moon formed between the stars long* before *the Sun was born and then was captured by the solar system.*" (*New York Times* September 21, 1969. Emphasis added.)

He adds: "How it then was drawn into a stable orbit around the earth would be difficult to explain."

Many scientists claim it's not just difficult but nearly impossible. NASA scientist Dr. Robert Jastrow admits that "the difficulty with that theory is that such a capture is extremely unlikely. For capture to occur, the Moon must have come by the earth at exactly the right distance, neither so far away that it was whipped past the earth without dropping into orbit, nor so close as to be drawn into a collision course. Calculations indicate that the range of approach distances that will lead to capture is very narrow, and that the probability of capture is exceedingly minute.

"Thus," admits Jastrow, "at the present time there is no adequate explanation for the evidence of the Moon as Earth's satellite." (*New York Times*, November 9, 1969.)

We note: The same thing is true today. Nothing can explain the mysteries of our Moon except the Soviet spaceship theory, which holds that our Moon (which is probably far older than our own star) formed out in the universe and was converted by unknown alien intelligences into a spaceship. At some indeterminate time in our past it was *steered* into orbit around our world.

For to assume that it could have done this without intelligent direction goes against all the laws of celestial mechanics, especially when one considers the Moon's nearly circular orbit.

For as Walter Sullivan concludes: "However, specialists in the movements of celestial bodies under the gravitational influence of one another—the science of celestial mechanics—found it hard to explain how the moon, if it came from afar and was captured by the Earth's gravity, achieved so well behaved and circular an orbit." (*New York Times*, November 16, 1969.)

HOW MOON SOIL IS OLDER THAN MOON ROCKS

We have seen the perplexing antiquity of Moon rocks. But another conundrum makes scientists even more puzzled. The soil in which the Moon rocks lay proved to be generally a billion years older than the oldest of the ancient lunar rocks. The contradictory age of the soil really threw Moon scientists into a lunar tailspin. Rocks taken from the Sea of Tranquility as old as they were proved nothing compared to the soil in which they rested, which proved to be at least 4.6 billion years old.

Other Apollo soil samples generally proved this lunar rule that the ancient soil was somehow older than the rocks. The Apollo 12 soil was a billion years older than the rocks that lay strewn in it. Scientists were absolutely bewildered, for this seemed to be utterly impossible. Scientists studying planets know that the soil is largely the powered remains of the rocks lying amid it. On the Moon, which for eons has been banged and bombarded by meteors and other heavenly missiles this should especially prove to be the case. To find rocks and pebbles on the Moon that were so old was perplexing enough, but to find that the soil in which they lay was approximately a billion years older really baffled scientists.

If this were not enough, our bewildered scientists were staggered by additional puzzling discoveries that followed in rapid succession. Lunar experts soon found out through chemical analysis that the lunar soil did not come from the rocks around it but from somewhere else. Where our poor confused lunar experts could only guess. In truth, they simply do not know.

At various yearly Lunar Science Conferences, which attract hundreds of scientists from all over the world not only as participants but as observers, this was one of the foremost topics of discussion. At these so-called Lunar Rock Festivals (as the scientists themselves dubbed the gatherings) researchers discussed various theories as to how lunar dirt and dust could be a billion years older than the rocks surrounding it. Where could it have come from? Says science reporter Richard Lewis: "In vain did one expert after another attempt explanations. . . ." In vain.

Unfortunately, to our knowledge orthodox straitjacketed scientists never openly considered the Soviet spaceship theory that the Moon was fashioned out of a natural asteroid and converted into a spaceship. If they did they could have understood how it could be a cosmic body that old. Also how it had to come from a much older section of the universe. But as it traveled through the cosmos, as Shelley puts it so strikingly, "among stars of different birth," clearly it would have passed through different cosmic "time zones," through areas much younger. In so doing it obviously picked up rocks and dust particles in the form of meteors and meteorites (including micrometeorites). And this could account for the rocks being younger than the ancient lunar soil itself. Moreover, it would solve another puzzling problem about the Moon. For although scientists found it nearly impossible to reconcile the great difference in the age of soil and Moon rocks, there also proved to be significant age differences among the rocks and rubble of the Moon itself. Scientists also were in a quandary over the age discrepancy of lunar rocks discovered lying side by side. As our astronauts made more and more trips to the Moon and took more and more samples, our Moon experts became more perplexed and mystified. Yet this too, in light of the Soviet theory, is understandable, for if our Spaceship

Moon did travel through the cosmos it would necessarily have picked up rocks and particles of widely different ages. This much is clear: The Soviet Spaceship Moon theory makes understandable yet another Moon mystery!

MYSTERY #2: THE MYSTIFYING CHEMICAL MAKE-UP OF THE MOON

What is the Moon made of? Certainly not green cheese, although Dr. O. L. Anderson now of the University of California (Los Angeles) reports striking similarities between the characteristics of lunar rocks and some cheeses. Strangely enough, astronauts also reported that at certain times under the pale greenish-blue earthlight, the surface of the Moon reflects a greenish hue.

One of the major findings of the Apollo explorations is that the Moon is "chemically different at different depths." Scientists have concluded that the Moon does not have the simplistic billiard-ball make-up that some scientists formerly conjectured it might. It is not homogeneous throughout.

Moreover, the composition is strikingly different from that of our Earth. From the more than a dozen different American and Soviet manned and unmanned explorations of Luna, we learned, much to our surprise, that our satellite is quite different in make-up. Elements have been discovered on her in great abundance which are extremely rare on Earth. Even more surprising, some elements found on the Moon have never previously been known to exist here in their natural forms. The Argone National Laboratory reported at the Third Scientific Conference in Houston that they had found uranium 236 and neptunium 237 in lunar samples taken by Apollo 12 and 14, elements never previously found before in nature!

A compound never before known to exist was a strange component called by Earth scientists "Kreep." It contains a high content of potassium, rare Earth elements, and phosphorous. This compound was discovered in Apollo 12 samples and dated at about 4.5 billion years old. The mysterious component is believed to be a part of the

Moon's ancient crust, although this contention has been challenged.

Moon scientists also announced the discovery of an entirely new mineral found in lunar samples. *Science News* reported (January 10, 1970) that this "unnamed mineral is a titanium-iron-zirconium silicate with concentrations of calcium and yttrium and lesser amounts of eight other elements including aluminum and sodium."

Many of these, as the reader now undoubtedly recognizes, are metallic elements with a very high melting point, those tough metals known as refractory elements.

We have already seen that scientists are puzzled as to the source of the intense heat required to melt and fuse these metals on the Moon. Although Earth scientists have not yet resolved this difficulty (except for the two Soviet scientists), it becomes clear that it was done through artificial means in the process of creating this cosmic Noah's Ark.

Dr. S. Ross Taylor, the geochemist in charge of chemical analysis for NASA, observed that it is strange that highly refractory metals should be found in such great abundance on the Moon, especially titanium, a metal used on Earth in building spaceships and supersonic jets. He called it a "fortuitous metal to find on the Moon"—a very apt metal for our Spaceship Moon. And this is an apt comparison, for if the Soviet researchers are correct in their theory that our Moon is really a hollowed-out spaceship then his remark is striking.

MYSTERY #3: WHY IS THE MOON SO DIFFERENT FROM EARTH AND SO DIFFERENT AT VARYING DEPTHS?

> *The Moon's composition is not at all what it should be had the Moon been formed in its present orbit around the earth.*
>
> —*Dr. Harold Urey,* Science News
> *October 4, 1972, p. 246.*

A close corollary of this second mystery is another mystery—"the moon differs from the Earth in a very strange way," notes Ubell.

Before man went to our Moon most scientists believed that our satellite was formed out of the same cosmic cloud as Earth, born with it about 4.6 billion years ago. It was therefore expected that its composition would be similar to that of our own planet. But again another big surprise: It isn't. Our lunar companion differs radically in make-up from Earth—and in a very mystifying way.

From our many unmanned and manned trips to the Moon scientists have gotten a fairly complete picture of our neighboring world, even though admittedly it is a very confusing one. Naturally, of course, since the lunar evidence gathered is limited, what we can learn about our Moon is limited. But then, we have difficulty in getting a complete understanding of the Earth under our feet, too.

Earl Ubell does point out: "Certainly none of our astronauts dug a hole 25 miles deep to determine the chemistry down there, but nature provided the shovel in the form of incoming meteorites that blasted very deep holes, and brought up the underlying rocks. Each of the Apollo landings, particularly Apollo 15, was designed to set down in areas where the dug-up rocks would have landed."

So we know a little more than our few trips' worth of rocks and data might indicate. Scientists, in fact, have learned enough about our neighbor to come to this conclusion, expressed by Dr. Don Anderson, professor of geophysics and director of the seismological laboratory at Cal Tech: "The Moon's composition is not all what it should be had the Moon been formed in its present orbit around the Earth." (*Physics Today*, March 1974.)

In general scientists like Anderson are puzzled not only by the differing elemental make-up of the Moon, but also over the question of why the Moon is so strangely chemically zoned, with its surface enriched in the refractory elements and its interior seemingly devoid of iron, the iron content seemingly plentiful in the maria and in one layer beneath the surface.

The abundance of the refractory elements like titanium in the surface areas is so pronounced that several geochemists proposed that refractory compounds were *brought* to the Moon's surface in great quantity in some unknown way. They don't know how, but that it was done cannot be questioned. These rich materials that are usually concentrated in the interior of a world are now on the outside.

In fact, the Moon is so strange in this respect that Dr. Anderson proposes that "the Moon was made inside out." (*Physics Today,* March 1974.)

We ask: Isn't what NASA scientists like Anderson concluded exactly what would be expected if the Soviet theory of a hollowed-out "inside-out" world were correct? We suggest that if we accept this Soviet spaceship theory of a hollowed-out planetoid, then the internal lunar materials —the insides, as it were—would have necessarily been brought to the surface during the process of hollowing it out, so that the Moon would naturally appear to have been made "inside out."

Furthermore, the fact that Luna is deficient in iron except for that strange layer under its outer crust and the iron-rich, titanium-rich maria again is understandable in light of the Soviet theory. For alien beings, the Soviet spaceship scientists maintain, used these materials in the formation of the spaceship's inner hull—that strange inner layer—and in flooding the Moon's outer crust, which they claim is really part of the outer patchwork shell.

The Moon, in short, appears to be an inside-out world because it is, in fact, a hollowed-out spaceship. This is just another piece of the puzzle that indicates that the planetoid circling us is nothing but a huge spacecraft.

MYSTERY #4: THE MYSTERY OF LUNAR RADIOACTIVITY

> *The upper eight miles of the Moon are radioactive with uranium, thorium and potassium . . . enough to produce a heart-breaking conundrum . . . How in the world did the radioactivity get to the top?*
> —Earl Ubell, "The Moon Is More of a Mystery Than Ever," The New York Times Magazine, *April 16, 1972*

Before man went to the Moon most Earth scientists thought that our satellite was formed cool, and it was believed by most scientists that it contained lesser amounts of radioactive substances than our Earth, its supposed sister planet.

However, even here there is a puzzler. For this is not the case at all. Ever since November 1971, when a thermometer placed by Apollo 15 astronauts turned up startlingly high temperatures scientists have been befuddled by the unusual heat problem. Extremely high readings here indicated that the temperature flow near the Apennine Mountains was very hot indeed. In fact, one scientist confessed: "When we saw that we said, 'My God, this place is about to melt! The core must be very hot.'"

This, it should be noted, merely indicated to scientists a high amount of radioactivity in this one area, and not necessarily a hot core or a hot Moon. Other measurements and data have since shown that the Moon's core is relatively cool.

Yet radioactive elements have been found to exist in the surface of the Moon in surprisingly great amounts. The amount of radioactive materials already measured in lunar samples is, as one NASA scientist put it, "embarrassingly high."

Even the early Apollo samples, such as those taken by Armstrong and Aldrin, revealed material containing uranium four times higher than in typical Earth rocks and fifteen times higher than in meteorites.* The bulk of material inside the Moon could not be this rich in radioactive uranium, for then the interior would be much hotter than is evident from the Moon's temperature.

In fact, if the Apollo 15 samples were representative of the Moon, enough heat would have been generated to melt the entire Moon. Similar data from the Apollo 17 site shows that it appears to be high in radioactivity also. It would seem to mean, then, that the Moon has far greater abundance of radioactive elements than our Earth. Why, scientists can only conjecture.

Many conclude that most of the Moon's radioactive elements must be concentrated in the upper regions. Otherwise it would be completely molten. In fact, scientists now estimate that the upper 8 miles of the Moon are radioactive with uranium, thorium, and potassium.

As Earl Ubell notes, "not so radioactive that it is dan-

* Rock samples collected from Sea of Storms has *twenty times* more uranium, thorium, and potassium than other lunar samples.

gerous to astronauts who trod the surface but enough to produce a head-breaking conundrum. How in the world did the radioactivity get to the top?"

Theories abound. Just to mention one: "In the earliest melting of the Moon the radioactive elements were carried upward with the slag. Need I add the idea has its fervent detractors?" writes Ubell.

This appears to be another solution to this mystery. In light of the Soviet spaceship theory there is an understandable answer to the conundrum, which, need I add, also will have its army of detractors.

If alien beings used intense amounts of radioactive elements to melt and hollow out interior portions of their Spaceship Moon, with the advanced technology obviously at their disposal they could have melted and poured out the interior onto the surface, using the proper mix of radioactive substances to generate sufficient heat to do this. This could thus account for the existence of such high amounts of radioactive elements found in the upper regions of the Moon.

Many lunar experts do believe that heat generated by radioactive substances (of course, in a natural way) melted much of the Moon's surface. However, another problem surfaces with the natural explanation, as many scientists have pointed out. Many top Moon scientists insist that melting of rocks through radioactivity is a *slow process and hardly could be responsible for the quick, massive melting* that obviously took place. This indicates it was not a natural process.

But, as we have observed, *artificial methods of an advanced technology could have used concentrated amounts of radioactive elements to produce this rapid melting that obviously was there.*

Maybe this is why so many radioactive substances and radioactive "hot spots" have been found on the Moon's surface, despite the fact that we have made only six trips to this strange world circling Earth.

Of course, this might be all sheer conjecture. But it is interesting to note that it does solve the problem and the puzzler as to how the radioactive elements did get to the top, and also makes understandable how radioactive substances could cause a quick, rapid melting. It also makes clear why these elements exist in such great amounts on the Moon.

The two Soviet spaceship theorists point out that they are convinced that the huge flat lunar plains of solidified lava were artificially produced by intense amounts of melted rock and mineral elements from the interior. It is interesting and important to note that according to *Science News* (April 7, 1973), "where elevation is high [highlands] the amount of radioactive material is low, where elevation is low, the amount of radioactive material is high. . . . This suggests to some that the radioactive rich basalts were a lava flow early in the Moon's history that filled in the lower regions of the Moon."

This suggests to us that it was used to cause such flows. Another lunar mystery that again proves understandable only in light of the Soviet spaceship theory!

MYSTERY #5: THE MYSTERY OF WATER ON A DRY, DRY WORLD

> . . . *a million times as dry as the Gobi Desert.*
> —Gerald Wasserburg

The first few trips to our Moon indicated to scientists that the Moon was a dry, dry world. Apollo 11, 12, and 14 had failed to find even the slightest trace of water on the surface of our neighbor. Also, the attenuated seismic responses led many scientists to claim that the Moon just had to be completely dry. NASA Assistant Director of Lunar Science Dr. Richard Allenby concluded: "There is no evidence in the rocks or geochemistry that water exists." In fact, the Moon was so dry that it prompted one scientist to make the observation that the Moon was a million times as dry as our Gobi Desert.

Then came Apollo 15 and the scientific world was rocked back on its heels again. A cloud of water vapor covering more than 100 square miles was discovered on the surface of the Moon.

Earl Ubell in his article "The Moon Is More of a Mystery Than Ever" says: "The Moon is dry, drier than the most dry of terrestrial deserts, and it doesn't seem to have ever had water in substantial amounts either on top or deep down. . . . Scientists were feeling very confident about the dryness when suddenly one of the instruments left on the Moon by the Apollo 15 broadcast back to earth on March 7, 1971, a signal that indicated a wind of water had swept across the Moon."

Where could a cloud of water vapor 100 square miles have come from? The two Rice University physicists who made the discovery, Dr. John Freeman, Jr., and Dr. H. Ken Hills, pored over the data and claimed that the water must have come from deep down inside the Moon.

Red-faced NASA scientists, however, first tried to pass off this cloud of water as coming from two water tanks of an Apollo descent stage left on the Moon. These two tiny tanks—60 to 100 pounds of water—NASA scientists tried to make the world believe created the 100-square-mile cloud!

The Rice University SIDE (Superthermal Ion Detectors) team noted that this could not be the case since the two detectors were located some 180 kilometers apart, yet the vapor was detected with the same flux at both sites. Moreover, they were pointed outward in the opposite directions, yet both had strong readings. Hence, concluded SIDE researchers, the water came from inside the Moon!

A few stubborn NASA scientists speculated that this 100-square-mile water vapor cloud might have come from the astronauts' urine which had been dumped into the lunar skies. This too was rejected by the SIDE scientists.

Although our first few astronaut teams on the Moon failed to bring back a shred of evidence that indicated the Moon had any water at all, our Apollo 16 astronauts did find rocks that contained rusted iron, and this indicated the presence of water somewhere on the Moon. For to have rust one must have oxygen and free hydrogen—this proves that there must be water on the Moon.

Scientists now admit that water exists there because evidence of hydrous material found on the Moon was confirmed by the University of Cambridge lunar science team. And B. J. Skinner of Yale University concluded: "The evidence seems good that we have the first hydrous mineral found on the Moon." (*Science News,* January 29, 1972.)

The water vapor problem may not seem too big a mystery from the standpoint of us water-world inhabitants. But its discovery absolutely astonished lunar experts. As the *Science News* states (October 23, 1971): "The presence or absence of water on the Moon is central to theories of lunar origin and development."

The impact of the water discovery was momentous. As lunar expert Dr. Farouk El Baz, then a leading NASA geologist, noted: "If water vapor is coming from the moon's interior this is serious. It means that there is a drastic distinction between the different phases in the lunar interior—that the interior is quite different from what we have seen on the surface." (*Science News,* October 23, 1971.)

How this can be perplexes scientists. But according to the Soviet hollow-Moon theory, our satellite has huge internal areas filled with gases for some kind of atmosphere to sustain life. Could these gases, which might include water vapor, be escaping through cracks and crevasses from the lunar interior out onto the surface of the Moon? The SIDE team scientists did conclude from their careful study that indeed the water vapor actually came from the Moon's deep interior. It would seem then that the water vapor clouds could have come from our hollow Spaceship Moon. In fact, there appears to be no other solution to the sudden discovery of this great cloud of water on this dry, dry world.

MYSTERY #6: THE MYSTERY OF THE MOON'S STRANGE MAGNETISM

We don't know what the magnetic story is but we do know that there are some very strange magnetic properties in these rocks which were not expected.
—Dr. Paul Gast

On the first Soviet and American flights to the Moon, tests indicated that it had little or no magnetic field. Then lunar rocks were discovered that were strongly magnetized. These rocks contained in effect "fossil magnetic fields" frozen into them, indicating that the Moon either once possessed or at least went through a global magnetic field of considerable strength.

This was a surprise to scientists, for the Moon was without a magnetic field—and it was assumed it had been without such a field in all its long existence. Then the discovery of the magnetized rocks.

The late great lunar expert Dr. Paul Gast, a geophysicist at Columbia University, observed: "We don't know what the magnetic story is, but we do know that there are some very strange magnetic properties in these rocks which are not expected." (Lunar Rock Conference, Smithsonian Institution, Washington, D.C., 1969.)

If, as the evidence indicates, the Moon did actually possess a strong magnetic field, this presents a bitter conundrum. For then it must have had a big iron core. But equally valid data and evidence show that a large hot core could never have existed inside the Moon.

What about the possibility that the Moon came close enough to Earth's magnetic fields to cause the Moon rocks to be magnetized?

As Ubell puts it: "If the moon were once near the earth so that the earth's magnetic field could magnetize moon rocks, the two bodies would have been so close that the moon would break up under the gravitational pull of the earth. And so on. Any ideas?"

Scientists seem to be caught here on the horns of a dilemma. Evidence that the Moon had or at one time entered a fairly strong magnetic field is certain. The origin of such a field is a mystery to scientists, for the Moon does not have a magnetic field now and scientists find it hard to believe it ever could have had one.

Then where did the magnetic field come from that left its telltale imprint on these rocks? Where did it go if it was the Moon's? Says Richard Lewis, "scientists calculated the magnetic field had to be as strong as 36 gamma field which is powerful."

Either that or the Moon had to pass through "extremely powerful solar magnetic or electric fields, which no longer exist." That presents the problem of why the fragile Moon did not break into bits, if it got that close to a body, such as Earth, emanating such a field. For to acquire the magnetism, it would have had to come that close. (*The Voyages of Apollo.*)

Here is where the Soviet spaceship theory again fits in so beautifully, resolving all these difficulties. With its strong internal hull, there would be no problems of being torn apart, even if the Moon did come close to another cosmic body. Such a celestial tug of war might cause the Moon to be lopsided (which it in fact is!), but its strong internal hull or shell would keep it intact. And, incidentally, the bulge on the far side of the Moon also indicates that the Moon does have tremendous internal strength and may have been in such a celestial tug of war at one time.

Did this happen when the Moon came near Earth? No one knows for sure. However, considering the age of Spaceship Moon, the opportunities for this to happen were many.

NASA scientist Paul Gast confessed before his death: "One of the exciting things about this paradox or enigma [of magnetism] is [that] perhaps behind all of this is an explanation that none of us are thinking about today. Eventually some smart person will sit down and have a bright idea which explains it all." (*Science News*, May 27, 1972.)

We are convinced that the Soviet scientists with their Spaceship Moon is the answer! For the artificial moon theory dissolves all these difficulties and, in fact, makes a lunar magnetic field not only plausible but likely.

MYSTERY #7: OUR HOT-COLD MOON—THE MASS OF CONTRADICTORY EVIDENCE

> *The puzzler here is that a cool moon contradicts other data that lead science to conclude that the Moon once had a molten core and a magnetic field.*
> —Earl Ubell, "The Moon Is More of a Mystery Than Ever," The New York Times Magazine, *April 16, 1972*

One of the greatest controversies revolving around the Moon before man went up to examine this strange world for himself was the question of whether the Moon was a hot or a cold body. A major question that seemed to grip lunar experts before Apollo was: Is the Moon, like Earth, a "living," internally hot body? Or is it, as Schopenhauer described it more than a century ago, "a frozen moon," cold and dead throughout?

The majority of scientists who had given this Moon matter any attention had come to look upon the Moon as a dead world, much as did the science-fiction writer H. G. Wells, who viewed it as a place of pockmarked craters, dead volcanoes, and lava wildernesses.

However, there were scientists like American astronomer Ralph Baldwin, whose intensive studies indicated to him that the Moon is a live hot ball. When Baldwin first undertook his pioneering studies, one of his professors said to him: "Why are you wasting your time on the Moon? It is dead and gone!"

But Baldwin's persistence convinced him and a large group of lunar scientists studying the Moon that this mysterious orb was in fact hot inside—or at least was at one time.

Nevertheless, most lunar experts remained skeptical cold-Mooners, despite the hundreds of transient lunar events, such as glows and lights seen on the Moon's surface, which other scientists tried to pass off as "volcanic activity." Some lunar experts, in fact, even openly scoffed at the idea that the Moon was *ever* hot enough to melt rock.

However, enough scientists were around who thought the Moon was still hot enough to possibly have active volcanoes so that some astronauts (Apollo 11) were even shown erupting volcanoes in Hawaii before they made their trip to Luna. Why? Because some scientists thought our astronauts might run into a few on the Moon!

On the other hand, a few scientists, like Dr. Harold Urey, doubted that there ever was any volcanic activity on our neighboring satellite. Urey claimed he could "show mathematically that the Moon was too small a body to have generated the kind of heat that would have resulted in lava flows covering entire maria." Observed Urey: "No geologist has yet been able to show me how you can get lava flows out of the Moon." (Henry Cooper's *Moon Rocks*.)

When his scientific opponents claimed that radioactive materials in rocks could have heated up the Moon enough to do this, Urey pointed out that this presents then another unanswerable problem. For if there had been sufficient radioactive-generated heat to melt rocks inside the Moon and start lava flows, then the Moon's interior should be hot today, as is Earth's. However, it is not.

Why should this be true? Astrophysicists claim that once a body the size of the Moon had heated up enough to melt its interior, it could not have cooled down within the time the Moon is supposed to have existed, since this 2000-mile-thick rock world is a pretty good "thermos ball," well insulated to keep its heat from dissipating. Thus it is difficult to explain its relatively cold condition today.

Unquestionably, however, there is evidence that our Moon was once very hot. It is everywhere on the Moon—especially in the maria, in its apparent huge volcanic lava flows. Even the first close-up American photos of the Moon showed to Baldwin and other scientists that the Moon was indeed a hot body, at least at one time. As Baldwin himself put it: "Surveyor [an early manned Moon probe] killed off the possibility of a cold moon!"

Later Apollo-obtained evidence of a once-hot moon was compelling. NASA scientist Dr. Gerald Wasserburg, pointed this out when he said "once [the Moon] got itself all hot and bothered and melted perhaps down to a depth of 100 kilometers." However, he added that scientists are unable to explain how it got that hot, when it did, and how it cooled off. (*New York Times,* January 12, 1971.)

But unquestionably, as Apollo results began to flow back to Earth, cold Mooners went into a decline. But the hot-Mooners were in trouble too.

As Earl Ubell noted in "The Moon Is More of a Mystery Than Ever": "The cold-mooners are in trouble because they have no way to heat up the Moon once the first melting, 4.6 billion years ago, had cooled down. [It was believed this was the beginning of the Moon as a hot ball.] The only way they can supply the heat for the subsequent melts is to place the radioactivity within the moon in just the right way so that it creates the melts at the right time. This, in turn, as Dr. Paul Gast of the Lunar Receiving Laboratory suggests, means that the moon was formed in different chemical layers instead of as a uniform mass."

But the hot-Mooners are in a dilemma trying to explain the evidence that shows the Moon is relatively cold today, since they have no theory to explain the mechanics that shut down this heating process.

Furthermore, the hot-Mooners are having difficulty trying to explain the other evidence that shows that the Moon is extremely rigid and cold on the inside—as, for example, the peculiar long virbrations that seismometers record after various impactings show, as does the existence of the strange bulge.

Yet another mystery is the high content of titanium found in rocks taken from the Sea of Tranquility—three times the amount found on Earth, in some instances more than ten times the amount!

The problem here is that scientists—even hot-Mooners—find it difficult to believe that the Moon was ever hot enough to have produced the melted rock that formed the titanium-rich maria. Even our own hot planet Earth is not that hot! Not a single scientist, not even the hot-Mooners has ever offered an adequate answer to this perplexing dilemma.

Today lunar experts are still strongly debating whether the Moon is hot or cold. Some data support the fact that it is hot; other evidence that scientists of the cold school can marshal tends to indicate that our satellite must be cold. Thus, the results of the various Apollo missions seem to contradict each other. One fact after another seems to cancel previously established facts and findings. No wonder today scientists find themselves in a confused quandary. No wonder Ubell states that "the moon is more of a mystery than ever."

We have already seen so many areas where this is true. We have seen hot-Mooners squirming over evidence that shows the Moon to be internally rigid which is indicated by its nonspherical shape and lopsided bulge. We have seen that the internal existence of mascons indicates that it is cold and rigid inside. And we have already noted that the peculiar vibrations that are transmitted throughout the interior over great distances and for long durations definitely indicate that it is rigid and cold in the interior.

On the other hand, cold-Mooners are having difficulty

explaining away the existence of once-strong magnetic fields, which many scientists claim required a molten-hot lunar core to create. Which, by the way, also leaves the hot-Mooners with the dilemma of getting the temperature inside down to where it is today—relatively cold.

As Dr. Paul Gast, chief planetary and Earth sciences specialist at NASA, who was a cold-Mooner asserted: "If it [the Moon] was at one time completely molten or near the melting point, then it is rather difficult to get it today, deep in the interior, down to the temperatures that [now] exist." Some would say that is impossible.

As Dr. Charles Sonnett, deputy director of NASA's Ames Research Center in Mountain View, California, concludes: "A core that is cool now almost precludes the Moon ever having a hot core, as the cooling mechanism is impossible to explain." (*Science News*, January 23, 1971.)

The core is relatively cool—even hot-Mooners now admit that. How then to resolve these contradictions and dilemmas?

Dr. Nowell Hinns of Bell Communications Inc. calls Dr. Sonnett's findings spectacular. The puzzler here is that a cool Moon contradicts other data that lead other, equally competent scientists to conclude that the Moon once had a molten core, once had a strong magnetic field, once was definitely hot. Where do we go from here?

To the Soviet theory! For it alone can make the contradictions and paradoxes understandable. It dissolves the impossibilities of these contradictions. If, as Soviet scientists Vasin and Shcherbakov maintain, some technology such as properly developed and placed radioactivity could make the Moon extremely hot and send oceans of lava to its surface, hollowing out its interior would have been easy. Then we would have a way of heating up the Moon to a temperature hot enough to melt elements like titanium; then we would have an explanation for all the evidence of volcanic melting on the surface of an essentially cold moon. Nor would a large hot core be necessary for this. In fact, some scientists believe that the tiny core the Moon might have once had was melted and poured out onto the surface, in part forming the maria which we now see as huge circular dark seas.

This would explain Urey's contradictory position of the impossibility of a cold Moon on the one hand and on the other, the irrefutable evidence of a once-hot Moon that Baldwin's evidence and the Apollo evidence indicate.

For as Ubell pointed out: "The only way they [the cold-Mooners] can supply the heat for the subsequent melts [after the first initial melt] is to place the radioactivity within the Moon in just the right way so that it creates the melts at the right time."

THE HEAT'S ON!

One difficulty that we have wrestled with is the heat of the deep interior that some scientists say exists. Although lunar experts claim for the most part that our Moon is internally a "relatively cool" body, some claim that at its core (which, amazingly, much evidence indicates does not exist!) it is still as hot as 1000 degrees. That is hot enough to melt lead. (Admittedly, our Earth is much hotter—somewhere between 3600 and 5000 degrees.)

Surely here is proof positive that the Moon could not be a spaceship?

Not necessarily, for, as we shall see, not only are scientists not sure it is hot but insist there is every evidence that it is, in fact, a cold body. Secondly, remember that if the Moon does have artificial construction on the inside and is a spaceship, all kinds of things are possible, from nuclear furnaces to a huge artificial sun that any such inside-out spaceship must have to lighten its dark interior.

But even aside from this, scientists are divided, some insisting that the Moon could not be anything but a cold body internally.

How did those scientists who are convinced that it is hot arrive at this temperature? How did they measure the temperature inside this chunk of rock 2100 miles thick and over 235,000 miles away from our planet?

Earl Ubell gives us a clear explanation in his article "The Moon Is More of a Mystery Than Ever," for he, too, tackles this problem of whether the Moon is a hot or a cold world.

He points out that every schoolboy (or schoolgirl, for that matter) knows from his grade-school science classes that a magnet placed under a piece of paper sprinkled with iron filings will arrange the iron particles into a pattern which shows the strength and direction of the magnetic field generated by the magnet. When you place another piece of iron near the magnet, that field will distort and the iron filings will change position and rearrange themselves. Actually, *any metal or other material will distort the field to some extent,* and by measuring the distortion scientists can determine the magnetic and electric properties of the disturbing body.

We have measured the magnetic fields around Earth and the Moon. Our Sun generates a strong magnetic field which envelops both our planet and our neighbor the Moon. Our lunar satellite Explorer 35 measured that magnetism. Our Apollo 12 astronauts also placed a device on the Moon which measured the Moon-made distortions in this magnetic field, and physicists were therefore able to obtain a good idea of the magnetic properties of the Moon.

What has this to do with the inside temperature of the Moon? Scientists have known that the electrical properties of a body vary in relation to its temperature. From these properties, they can get a good indication of the temperature of that body. In fact, scientists have actually worked out mathematical formulas based on this relationship. Experts using these formulas have determined that the *maximum* internal temperatures of the Moon are between 1000 and 1800° F. This is the *maximum*. It could be much cooler, and many scientists insist it must be.

However, as we have just noted, any metal or other similar material will distort the field. If this is true, then we can see we will have problems with the accuracy of heat measurements of our Moon, since it is loaded with metals, including a complete metallic shell on the inside! For, as our Soviet scientists, theorize, this Spaceship Moon has an inner hull created by alien beings to protect themselves in their cosmic journey through the universe. We have seen there is strong evidence today that this iron does exist, in addition to many other concentrations of metal; there is the very real possibility of artificial construction—like the huge girderlike blocks of metal 1000 kilometers long that may exist inside our Moon! Thus, clearly the results of any temperature studies would be highly inaccurate, to say the least.

In addition, the temperature readings have been taken in only a few spots on the Moon. Our first attempt to place heat-temperature measuring equipment ended in failure when on the Apollo 15 expedition David Scott unfortunately stumbled, ripping out the wires of this sensitive heat temperature equipment.

However, Scott and his partner James Irwin did succeed (after many tries) in drilling a hole in the lunar surface and placing in it a thermometer.

In November 1971, the thermometer readings indicated that the heat flow in that area (near the lunar Apennine Mountains) was *two thirds the Earth's average*! For an orb as small as the Moon, that is very hot! As one scientist exclaimed: "When we saw that we said, 'My God, this place is about to melt! The core must be very hot.'"

However, this, as the scientists themselves well knew, was an isolated reading. And what evidence do scientists have today for determining the Moon's internal heat? A few isolated readings. We have noted the large number of radioactive hot spots already found on or near the surface of this strange world. Considering this, it is doubtful that these few readings are very accurate. NASA clearly admits that many hot spots have been found on the Moon (just as on Earth) where there is a much higher heat flow than the average. (Yellowstone National Park is a good example for Earth.) So any estimate of the inside temperature of Luna based on this is hazardously uncertain.

Dr. Michael Yates of Bell Comm Inc. claims: "The magnetism results from Apollo 12 seem to be supplanting the seismic results in importance since the magnetism relates to the Moon as a whole." (*Science News,* January 23, 1971.)

Although he believes that findings do show that temperatures increase to a depth of about 200 kilometers, he also is convinced that they *decrease* again. From this he concludes that the Moon is a relatively cool body, certainly well below the melting point of most solids.

We have also seen the seismic studies show the Moon vibrating throughout like a huge bell, and that this indicates by itself that the Moon interiorly is cold and rigid. Similarly, deep seismic disturbances show the Moon to be rigid and cold. The highly regarded *Science* magazine notes: "Possible focal depths as great as 900 km. may favor the thermal stress hypothesis, but there is considerable evidence that melting temperatures *may not be reached at any depth in the moon*. Relatively low internal temperatures are indicated by the existence of mascons and by the non-equilibrium figure of the moon which imply considerable internal strength." (Emphasis added.)

There is also much other evidence that our Moon is internally not hot but a cold orb. Many studies have led to this conclusion. Here are a few:

English scientist C. A. Cross published his study "The Thermal History of the Moon" in the authoritative international lunar journal *The Moon*. He says ". . . for various reasons the outward flow of heat must be smaller than is presently supposed." (*The Moon, An International Journal of Lunar Studies*, Vol. 4, 1972.)

Dr. Marcus Langseth of Columbia's Lamont-Doherty Geological Institute also warns that for various reasons "we cannot extrapolate the temperature increases with depth." Dr. Langseth also claims that those few high heat flow measurements that have been taken "could be a local anomaly." (*The Moon*, Vol. 6, 1973.)

We have seen how much intense radioactivity permeates the upper 8 miles of the Moon, with hot spots everywhere. This too could be affecting the evidence for a hot Moon.

NASA scientists Dr. D. L. Anderson and T. C. Hanks point out in their article "Is the Moon Hot or Cold?" that there are many problems with any theory of a hot moon, among these the non-spherical shape of the Moon, the mascons, the remarkable seismic responses and conductivity evidence! All suggest the lunar interior is "cold and, by implication has always been cold." (*Science*, Vol. 178, 1972.)

Nobel prize winner Dr. Harold Urey stated bluntly: "The geologists are going to be in for an awful jolt when they find the Moon isn't volcanic in the way the Earth is.

I keep telling these geologists to let their imaginations go!" (Henry Cooper's *Moon Rocks*.)

Little did Dr. Urey realize the full impact of his prediction. Henry Cooper in his fine book *Moon Rocks* commenting on Urey's remarks says: "A hot moon would have to have had a hot core." Did it? In the orthodox view, yes. The problem with this is that equally valid data indicate that the Moon could never have had a hot core—at least not naturally.

And this is the key word—*naturally*. How can you have a hot core when evidence indicates that the Moon could never have had a hot core *naturally?* But Vasin and Shcherbakov maintain that it could have been done *artificially* through advanced alien technology's reshaping of this natural world into an artificially created spaceship.

So maybe Urey's old mathematical proof is not wrong after all—maybe the Moon is too small a body to have generated the kind of heat that would have resulted in lava flows covering entire maria. Maybe the Moon was heated up artificially with advanced technological means by intelligent beings to accomplish what appears impossible to Earth scientists. This, as we have seen, undoubtedly involved large amount of radioactive elements placed in proper fashion within the Moon.

Technologically advanced aliens could have "heated" the Moon, sending seas of lava to the exterior surface initially as they hollowed out inner portions and later on at various times to reinforce certain weakened outer areas. Thus, since the Moon had no hot internal core, there exists no problem of trying to figure out how it cooled.

This spaceship theory would allow the Moon to preserve its lopsidedness or aspherical shape as well as the mascons. It would also explain why the Moon gives every evidence of being cold and rigid in addition to being hollow internally. Yet with this artificial process it still could have hot spots in its interior and in its outer regions, which seemingly are merely intense concentrations of the radioactivity used by the aliens to refashion this natural asteroid into a spaceship.

This theory thus explains all the seeming mysteries, which in fact melt away in the heat of the facts of this "artificial" hypothesis. Suddenly everything becomes clear in the light of the Soviet Spaceship Moon.

> . . . *although the Moon has been visited several times, no answer has been obtained* [as to its origin]. *The information we have received offers more puzzles than answers.*
>
> —*Isaac Asimov*

THIRTEEN
THE ALL-COMPREHENSIVE SPACESHIP MOON THEORY!

Ever since the Apollo missions started bringing a new world of information to Earth our scientists have used this new-found lunar knowledge to kill off many of the most widely accepted theories of the origin and make-up of the Moon. Scientists began dispatching rival theories left and right because they did not completely fit the lunar evidence as a whole. At the same time scrambling scientists are creating new hypotheses to take the place of old, worn-out ones. These new theories are being fitted Procrustean-like to the strange discoveries made in exploring this strange world.

"A new Moon rose," Richard Lewis told us at the Fifth Lunar Conference in 1974. "It was a planet in its own right . . . such a planet had to be captured to become the satellite of the Earth." (*The Voyages of Apollo*.)

This is the latest theory, now being advanced cautiously to try to explain some of the enigmas of the old Moon now that most of the other theories are dead or dying. And old scientific theories, believe me, die a very slow death.

Interestingly, the theory that is now gaining most adherents is one that was never seriously considered by the vast majority of scientists before Apollo. Most experts in the field of celestial mechanics rejected this possibility. As Walter Sullivan, a leading science writer of our time, says:

> . . . specialists in the movements of celestial bodies under the gravitational influence of one another—the science of celestial mechanics—find it hard to explain

how the moon, if it came from afar and was captured by the earth's gravity, achieved so well behaved and circular an orbit.

Even with all their fancy computers, their fine spacecraft, guidance and velocity control, specialists in space flight find it virtually impossible to launch a vehicle from earth so that, without further nudging, it goes into orbit around another body such as the moon. As in all the Apollo missions including this one, a rocket engine must be fired to inject the spacecraft into orbit as it flies past the moon.

Obviously the moon had no such rocket, so how did it get into orbit around the earth? (*New York Times,* November 9, 1969.)

A good question, and one that only the spaceship theory of the Moon's origin resolves with any intelligence. The science of celestial mechanics and its laws have not changed. The odds of a natural capture of our Moon, considering its nearly circular orbit, are virtually nil. It is virtually impossible.

Another insoluble problem is that in order for an orb the size of the Moon to be captured by Earth it would have had to come within the Roche limit—that is, within that limit which would cause it to be pulled apart and break up under the tremendous effect of the great gravitational pull of our much larger, denser Earth. Only a Spaceship Moon with a thick metallic hull could have withstood such a celestial tug of war and survived intact. Even at that, the smaller Moon would have been left with a huge bulge on one side, at the very least. Interesting, isn't it, that our Moon has such a bulge—on the far side, which has never faced Earth.

A NEW STRANGE MIRACULOUS MOON

At the Fifth Lunar Conference scientists like John Wood and Joseph Smith of the University of Chicago and H. E. Miller of the Smithsonian Astrophysical Observatory maintained that the Moon *did break up* as it passed our planet on its way to being captured. Then, according to their theory, the Moon "reassembled itself in earth orbit." Amazing. They do not exactly spell out how this miracle took place, except to say by accretion, an old standby of astronomers.

Incidentally, these theorists use this "capture-breakup-reassembly" theory to help explain how the Moon lost its iron, the element that, as we have seen, is lacking in the outer portions of the Moon, except of course in the maria, where it is strangely concentrated. They claim that the heavier iron pieces of "the disrupted invader were dispersed in distant earth orbits." The remaining pieces, ranging in size from tiny pebbles to huge meteor-size boulders many miles across, were drawn to the Moon by its gravity and crashed into the ever-enlarging Moon, until it eventually generated enough heat to turn the lunar surface into a sea of molten lava. This rain of rocks would have to have happened very, very rapidly, which scientists are at a loss to explain. These theorists assume also that the central core of the Moon did exist, because they start with that and use it to build up to a large Moon.

This rain of rocks ended some 3.9 billion years ago, leaving the Moon covered with craters and maria. Radioactive elements like uranium and thorium gradually built up, melting once solid interior rock, finally breaking through the hard outer crust in great spasms of volcanic activity, filling the Moon's low lying basin with lava, creating the Moon's maria.

This is hard to accept, because most scientists point out that radioactivity would not melt enough rocks quickly enough to send forth oceans of lava larger than entire countries of Earth. Furthermore, there is much evidence, as we have noted, against the celestial-bombardment-formation theory of these strange circular seas of metallic lava.

Moreover, if the Moon was indeed formed from huge chunks of matter in orbit around Earth, where did the debris come from? Scientists who espouse this wild theory have an ingenious answer to this problem. Geochemist John Wood of the Smithsonian Astrophysical Observatory in Cambridge, Massachusetts, suggested that thousands of chunks of this rock matter were shooting through this corner of space, "trapped near the earth and broken up by gravitational forces, but many of their heavier components, notably iron, were thrown back into space. That would explain the paucity of such material in moon rocks." (*Science*, 1973, Vol. 103. Also see *Time*, April 8, 1974.)

Convenient, convenient. Why would the heavier components, notably the iron, be thrown back into space when by the laws of gravity they should have been attracted to the surface of the Moon even more quickly and surely than lighter chunks?

And how to explain by this theory why Earth, which is larger to begin with and a close companion now, did not attract these large chunks of lunar debris? Certainly, our planet in no way shows signs of going through any kind of intense bombardment; the Moon does.

No, this theory appears to us to be quite contrived. It seems more unlikely an hypothesis than the far-out Soviet theory of an intelligently created spaceship world steered into orbit around Earth. More importantly, this "new Moon" still leaves us with many unsolved problems—with the mysteries of the mascons, the contradictory cold-hot evidence, the mystery of how the Moon could ring like a huge gong for over four hours, the mystery of identical signal tracking from inside the Moon, and myriads more!

As far as we can see the best that can be said for it is that it shows how lunar scientists are now leaning toward the probability that the Moon was born far from our planet and came from the outer cosmos. The chemical differences alone are leading some scientists in this direction. For if Earth and the Moon were created out of the same cosmic cloud, why do they differ so much in composition? These differences cause many scientists today to discard the theory that the Moon came into being at the same time as Earth. The theory that the Moon was ripped out of Earth now seems to be completely dead and buried.

The capture theory is now catching on. And that's good, because it is a step in the right direction—toward the Spaceship Moon theory! As Gustaf Arrenhenius and H. Alfren of the University of California claim: "There is little doubt that the moon is a captured satellite. But its capture orbit and tidal evolution are still incomprehensible." (*The Moon, An International Journal of Lunar Studies*, Vol. 5.)

And will remain so until, as we have pointed out, someone can find some way to overcome the problems of celestial mechanics arguing against natural capture.

There are many other grave problems with the "new" Moon of Wood and Smith and their colleagues. Just how, for instance, did this Moon by accretion "reassemble" itself so that the iron existed in intense amounts only in the maria and in an inner band or layer just below the Moon's rock surface? They do not—in fact, cannot—explain this conundrum. Obviously, this theory seems to be an unlikely explanation even compared to the wild Soviet theory which holds it was extracted by alien intelligence to construct the inner portions of the spaceship hull and reinforce the outer weakened areas of the Moon's shell.

After that miracle of nature flinging the iron out of the Moon, these scientists of the "new" Moon claim, the major pieces recombined near the Earth and minor pieces were picked up later.

Maybe if scientists would take a close scientific look at the inner Moon they could see where all the iron and metal went. In fact, scientists, as we have shown, know that a thick inner layer is rich in iron and refractory alloys. And we have shown one outstanding lunar expert trying to explain the strange 1000-kilometer-long belts of seismic disturbances by saying that they could be "composed of material such as *embedded blocks of iron . . .*"

In our opinion, in no way could iron blocks exist naturally inside the Moon. Except, naturally, in the unnatural way of the artificial construction of a spaceship!

Scientists, as you can see, are practically doing lunar handstands trying to figure out how, as one science reporter put it, "the moon accreted with less iron, more aluminum and other refractory elements and more uranium than the earth. 'It ain't easy,' quips one geochemist." (*Science News*, April 7, 1973.)

Anyone can see this. And in fact, to acceptance of this entire hard-to-swallow capture-breakup-reassembly we reiterate that geochemist's quip: "It ain't easy!"

We would rather stick to the wild, way-out, lunatic-fringe spaceship theory that fits all the facts. The capture-breakup-accretion-reassembly theory doesn't.

As Dr. Donald Wise of the University of Massachusetts,

a proponent of the fission theory, once said: "It's always possible to choose among competing theories if you're willing to disregard half the evidence." (Henry Cooper's *Moon Rocks*.) This strange theory is guilty of just that, as we have seen.

In fact, scientists have yet to come up with a unitary theory that does fit all the facts—aside from this unorthodox Soviet theory. We have seen how this amazing theory dissolves all the mysteries of the Moon. And, poring over the Apollo evidence, we have yet to find a single problem or contradictory piece of evidence that is not understandable in light of this artificial-Moon theory.

THE MYSTERY OF OUR GLASSY MOON

Close-up photos taken by our lunar astronauts show the Moon's surface looks as if it were covered with solder. On close scrutiny this metallic-looking "solder glaze" turns out to have much more of a glassy effect than a metallic one, scientists discovered. It appears that the Moon's hide was scorched somehow by some intense heat. What caused this strange glazing phenomenon?

At first some scientists were convinced that the glassy beads were merely condensed out of surface material burned by the impact of meteors.

In fact, one science reporter wrote: "The discovery that some parts of the Moon are 'paved' with pieces of glass supports the view that the Moon has suffered impacts of a very energetic nature." (Frank Cousins, *The Solar System*.)

Did all this glass come from massive meteor impactings?

"Nope," says eminent NASA scientist Dr. Eugene Shoemaker. "Zero chance.

"If the surface material vaporized, gas molecules would have so much energy that they would escape into space rather than condensing and falling back onto the surface in such large numbers." (*Science News*, August 2, 1969.)

Where did the glass come from, then?

Dr. Thomas Gold, leading astronomer of Cornell and NASA programs, said that this glass suggested to him that it came from the scorching of some intense heat, probably produced by solar flare-ups. Calculations, however, show that the Sun would have to increase its luminosity by more than 100 times its present rate for at least 10 seconds and probably up to 100 seconds to accomplish this. Such a phenomenon would have made it a kind of mini-nova. Most astronomers discount such an intense flare-up by our Sun because it appears to be such a stable star and because there is no evidence on Earth of such a great solar scorching.

Dr. Gold insists that the evidence indicates clearly that the flash of intense heat that scorched the Moon's surface must have been recent, because otherwise the erosive effect of the constant cosmic rainstorm pelting the planetoid would likely have obliterated this lunar evidence. Gold estimates that this intense glazing could not be older than 100,000 years, and more likely closer to 30,000 years.

Scientists are stumped to explain another Moon mystery. But in light of Spaceship Moon we do not need a mini-nova of our Sun to explain it. If this Spaceship Moon traveled throughout the universe for eons, there would most likely have been many occasions when it could have occurred. It undoubtedly zipped close to many another sun, so it would have had many opportunities to get such a fierce scorching.

Whether this is what happened we cannot say, but certainly this problem disappears in view of the travels of our Spaceship Moon.

OTHER EVIDENCE OF OUR WANDERING MOON

Yet another piece of evidence whose understanding is important to the entire picture of the Spaceship Moon puzzle comes to us through our latest Apollo studies.

Dr. Z. Kopal of the University of Manchester, who is looked upon as one of the world's leading experts on our lunar companion, in his definitive work *The Moon in the Post-Apollo Era* zeroes in on another problem of the Moon that many other Moon experts overlook.

This is the conclusion that a study of lunar craters reveals the "fact that evidence for impacts in all directions—including low-angle or grazing impacts—appears to be distributed at random all over the surface of our satellite."

Kopal's inquiry into the orbits of the impacting bodies

that formed the craters and the trajectories that they must have followed leads him to the unalterable conclusion that these celestial particles or pieces which hit to form our Moon's craters were *not in heliocentric orbits.*

As Kopal stresses: "The stony record of the lunar face" indicates that "the impinging particles could not have revolved in heliocentric orbits." He arrived at this hard and firm conclusion from a study of the craters and the angles of impacts, taking into consideration the known laws of celestial mechanics.

Some scientists have considered that cratering on the Moon took place while the Moon was in orbit around our Earth. This presents a ticklish problem. Kopal now claims that convincing evidence indicates that such cratering could not have taken place while the Moon was locked in orbit around our Earth and Sun.

"Is there any escape from these conclusions?" asks Kopal. "One would be to assume that for the most part the continental impact craters were inflicted on the Moon *before its eventual capture by the Earth*—when the Moon, *travelling alone through space,* may have been tumbling [spinning] about its centre of mass, so that impacts from all directions would have been equally likely." (Emphasis added.)

Again, the evidence of the Moon clearly points to the fact that our Moon must at one time have been a wandering world, journeying through the universe. But could not the Moon have done this naturally, then accidentally been captured by Earth's gravitational field and locked into orbit around our planet? We have seen that this unlikely capture theory presents many difficulties, not the least of which (as we have pointed out time and again) is how the Moon could have been captured (and many scientists doubt that it could have been) and yet still have taken up its circular, well-behaved orbit around our world. Also, the peculiar position the Moon has assumed around our world, which permits it to produce eclipses, does indicate, as the Soviet scientists suggest, that it was powered and steered here!

But this is not all. There is other convincing evidence that this is exactly the case. For Kopal adds in conclusion: "We have, therefore, no idea whether or not the Moon could have been spin-stabilized in space before its hypothetical capture. However, even if this was not the case, it is most unlikely that any of its pre-existing surface sculpture would have survived so brutal an experience as capture of the Moon by the Earth would inevitably have been. The need to slow down a passer-by to make capture possible would have required a dissipation of so much energy of the Moon through inelastic tides as could be raised only at a very close approach; and such tides would no doubt have obliterated most, if not all, of the pre-existing surface-markings."

But they did not! Clearly this is another indication that this was not an ordinary "capture." With this Spaceship Moon, of course, it was really no capture at all. It was, in fact, an intelligently directed, powered, and steered maneuver that put the Moon in orbit around our Earth—so that in a sense the Moon captured Earth and not the other way around. They adopted us; not we them.

Another difficult problem solved in light of the Spaceship Moon theory!

This is the beauty of the Soviet spaceship theory. It is comprehensive! It answers all the problems and puzzles; solves all the mysteries and enigmas, dissolves all the lunar contradictions and difficulties. It is the perfect solution, for it fits all the evidence! And that evidence, as we have seen, turns out to be proof that our Moon is a hollowed-out spacecraft, artificially transformed from a ball of rock into a metallic inside-out world.

FOR ONLY THE "SPACESHIP" MOON MAKES UNDERSTANDABLE:

1. Why the Moon is a freak world—too big and too far out to be the natural satellite of Earth.
2. Why the Moon could have a synchronized, almost perfectly circular orbit.
3. Why the craters of the Moon are so numerous and so strangely shallow.
4. Why the Moon could have a great bulge on its far side, which never faces Earth.
5. Why the Moon has such great internal strength.
6. Why the Moon, so bereft of iron in general, has a rich band of iron and other metal in an inside layer.

7. Why some Moon rocks are older than Earth and even our solar system!
8. Why lunar rocks lying near each other can have such varying ages.
9. Why lunar soil can be a billion years older in general than rocks lying about it.
10. Why the Moon's composition is so different from Earth's.
11. Why the Moon seems to have been made "inside out."
12. Why the maria and mascons can exist.
13. Why the maria have so much metal content of a highly refractory nature.
14. Why the Moon can have pure metal particles, including iron that does not rust!
15. Why the Moon has heavier material that flowed to the top, apparently against the laws of nature.
16. Why the Moon has evidence supporting both hot-Mooners and cold-Mooners, indicating the Moon must have been hot and at the same time proving it never could have been.
17. Why the Moon has evidence of tremendous heating and melting and yet is relatively cool today.
18. Why the Moon vibrates like a huge gong, conveying tremors great distances and even completely around the Moon.
19. Why the Moon appears to be a huge rubble pile.
20. Why the Moon has the "shakes" periodically, with swarms of seismic activity.
21. Why the Moon seems to have such low density and peculiar lightness, in that it gives every evidence of being partly or wholly hollow.
22. Why seismic waves travel so fast through an extremely hard inner layer, indicating a metal shell under its rock crust.
23. Why the Moon produces identical seismic waves of internal disturbances.
24. Why the Moon can be so precisely in the position it is so that viewed from Earth it is equal in size to the Sun's disk, each canceling the other out during eclipses, thus making possible such a unique phenomenon—the only planet we know of that has this phenomenon.

In addition to all this, the spaceship theory also makes understandable such lunar enigmas as:

1. How the Moon could have been captured by Earth or should we say, how the Moon captured the Earth, yet ended up with its unexplainable circular orbit.
2. How the Moon can be where it is, not in the usual satellite orbit around our orb's equator, but instead following strangely an orbit closer to the Earth's own orbit around our Sun.
3. How the Moon can sustain its great lopsidedness.
4. How the Moon could have melted on its surface such elements as titanium and zirconium that have tremendously high melting points.
5. How the Moon can have so much radioactivity in its upper layers.
6. How the Moon can have internal vibrations and tremors that last for hours.
7. How the Moon can have internal disturbances at unheard of depths.
8. How the Moon can affect the magnetic needle of a compass, even if ever so slightly.
9. How the Moon can have indications of being extremely rigid at great depths and yet seemingly have other properties and evidence to indicate it is warm.
10. How the Moon's outer crustal surface could have been melted so rapidly.
11. How can there be so much metal in the Moon's maria, even pure processed metal!
12. How and by what forces the Moon could have had such significant redistribution of its crustal materials.
13. How the Moon could have come to have a strong remanence without a molten core.
14. How the Moon can be a dry-as-dust world and yet have occasional huge clouds of water vapor.
15. How the Moon can be volcanically dead and yet have had hundreds of strange glows and lights and moving objects on its surface in the past several centuries.
16. How the Moon can be a hollow shell yet not collapse (due to its great metallic shell).

17. How the Moon can spawn so many seeming natural contradictions of data and findings.

BECAUSE IT IS, IN FACT, AN INTERNALLY, ARTIFICIALLY TRANSFORMED WORLD—A SPACESHIP WORLD!

UFOs are astronautical craft, or entities. If they have a fixed base of any kind, that base is likely the Moon.
—Morris Jessup

FOURTEEN
SOMEBODY IS INSIDE THE MOON!

We have seen how remarkably the American and Soviet evidence obtained from Moon expeditions all seems to back the startling spaceship theory.

Admittedly, there are a lot of loose ends and unanswered questions. One staggering one: *Is the Moon an island of life or a graveyard of ages past? Is it a celestial Flying Dutchman floating through space that once harbored alien beings or is it still teeming with alien life? Are there intelligent beings there today?*

We have seen in our opening chapters that hundreds of weird lights and moving objects seen on the Moon indicate that it may indeed yet be an island of life, either a home of aliens or a base for visiting beings. Hundreds of strange sightings in the past several centuries since the invention of the telescope convince us that such may be the case. And the strange UFO encounters our lunar astronauts had are the clincher, the final proof.

Admittedly, the Soviet scientists who devised the spaceship theory speculate that Spaceship Moon is an abandoned world. But, as we have seen, it may not be a completely dead celestial body after all. Maybe it is not alive volcanically, but there is definite evidence that it is alive biologically—that is, there are live alien beings up there (living, of course, on the inside).

Strangely enough, when the Apollo Moon program began, ostensibly the orthodox scientists of Earth were only looking for simple forms of life, like microbes.

Did lunar scientists expect to find life on the surface of this seemingly dead world? When Dr. Eugene Shoemaker, one of NASA's leading scientists, was asked this question he laughed out loud.

"Well, of course, there are the lunar elephants!" he responded. "We'll see those!"

Shoemaker was referring to a pompous seventeenth-century English researcher, Sir Paul Neal, who announced after peering through a telescope that he had discovered an elephant on the Moon! Needless to say, this announcement caused quite a stir of excitement until, as the story goes, people found that all he had seen was a mouse that had apparently crept into his telescope.

LIFE ON THE MOON?

One problem foremost in the minds of many Earth scientists was the task of landing astronauts on the Moon without contaminating the surface with earth-bred organisms and germs. And vice versa, without contaminating Earth upon the astronauts' return. This theoretically could be fraught with great danger. So strict precautions were taken. Our first astronauts were isolated and examined for many days upon their return to make sure they had no contaminating microbes on them.

Scientists were intensely interested in finding out whether our Moon had any evidence of life on its seemingly sterile surface. The dust brought back to Earth from the Moon seemed to be as dead and bereft of life as the Moon itself appears to the ordinary observer.

No evidence of life was found in any of the Moon rocks or soil brought back to Earth, although one physicist examining a sample of lunar dust was startled when he looked through a microscope to see what appeared to be a Moon organism with legs and feelers! It turned out to be only a flea that had come from a dog which had strayed into the lab.

The precious Moon dust was examined microscopically very carefully and subjected to over 300 different tests to see if anything would live in it. Nothing did. One scientist thought he had found what appeared to be columns of microbes. However, the pretty turquoise mold turned out to be a wholly chemical phenomenon.

Although no Moon organism was found growing in 300 different environments, strangely enough, scientists found that Moon dust itself seems to help plants to flourish and grow lustily.

Scientists, many of whom were from the U.S. Department of Agriculture, were unable to account fully for this phenomenon. Nevertheless, some were convinced after several experiments that it had to do with the chemistry of the lunar dust itself, which appeared to be much like volcanic ash of Earth in the way it stimulated growth. But in essence, chemists and geochemists alike were mystified by this puzzling phenomenon.

In any event, the important discovery does show that plants can be grown on Moon dust, which means that a Moon colony, if ever established, would have no difficulty in furnishing sufficient food from luxurious lunar gardens. For it seems to be an ideal medium to grow lush gardens in.

ORGANIC COMPOUNDS FOUND ON THE MOON!

The Moon has been labeled by most scientists a dead world, volcanically and biologically. Seemingly, no life exists on its surface. And no direct evidence of life has turned up in its dust and dirt.

And this seems to be just what most scientists expected. For this waterless, airless blasted rock in our skies is full of deadly radiation and racked by tremendous changes from day to night, which of course does not make it a likely haven for any kind of life. On the outside it appears to be one huge sterilizer in the sky. But, clearly, who knows what exists underneath the protective layers of rock and crust? Life might now be thriving in the interior of this mystery world.

It is interesting to note that in mid-1971 a team of biochemists announced that they had found traces of organic chemicals in a sample of soil returned from the Moon. "They found complex carbon-based compounds—not living matter, or even material that was alive. But the long chain, carbon-based molecules that you and I are composed of. If organic compounds have been found on the Moon, can there be life there too?" asks science writer Ben Bova. (*The New Astronomies,* St. Martin's Press, 1972.)

Similarly, the scholarly periodical *Scientific American,* in an article entitled "The Carbon Chemistry of the Moon" (October 1972) claims that "exhaustive analysis of the Apollo samples reveals various simple organic compounds. These substances did not originate with life, but they add to the store of information on how life originated."

And we add with *Scientific American* that they make it "increasingly likely that we are not alone in the universe."

However, this highly respected journal of scientific thought appears to be ignoring the overwhelming weight of evidence that not only is our Moon a spaceship but it is an occupied spacecraft! For the facts show the Moon is teeming with intelligent life!

INTELLIGENT LIFE ON THE MOON

Not only have strange lights, weird moving glows, and unidentified Flying Objects been spotted on the Moon throughout the telescopic age of man—the period man has been observing our satellite through a telescope—but other strange phenomena taking place constantly on her surface indicate that life is indigenous to the Moon. That it has been there a long time and appears to be using the Moon as a base for operations.

Perhaps the most unexplainable reports of all were the astronomic sightings of strange structures and weird changes taking place on the surface of this supposedly dead and changeless world.

Astronomers have concluded that the Moon is an airless, windless, and for the most part erosionless world. At least, this is what scientists have been telling mankind for a long time. It is a changeless orb. Yet in the past several centuries many peculiar changes and happenings have been detected by competent observers, taking place on the Moon's surface right under their eyes, for which they have no satisfactory explanation.

We have seen how expert astronomers like the late Morris Jessup, formerly a leading astronomer of the University of Michigan, concluded that these lights and bright moving "spots" astronomic observers all over the world have been reporting over the past several centuries are nothing else than UFOs—spaceships of alien beings on the Moon!

And indeed many of these strange reported sightings do seem to be just that. For instance, on September 7, 1800, several people in France observed strange objects moving in straight lines above the surface of the Moon during a lunar eclipse. They appeared to be close to the Moon's surface. Then, as these observers watched, the objects all turned in the same straight line, separated by uniform spacing, as precise as though they were in military formation.

In his book *The Case for the UFO* Jessup concludes also that in addition to lunar lights there are exterior dome-like constructions on the Moon that indicate their presence there.

Evidence that they have been building strange domelike structures on the Moon surface is impressive. Several hundred have been reported. Some claim that they now number over 700. Points out astronomer Morris Jessup: "Two hundred years ago there were none." (*The Expanding Case for the UFO*.)

Some of these strangest of all lunar structures are rather large—nearly 700 feet in diameter. The great English astronomer Dr. H. P. Wilkins points to one discovered as late as 1953. On September 26 of that year, F. H. Thornton sighted "a sort of island of light amid the blackness, clearly proving that it was raised, in fact a dome." (*Our Moon*.)

Wilkins observes: "It is curious that this is the first time that such a thing has been seen. Is it possible that it has only become a dome recently." (*Our Moon*.)

This great English astronomer points out that earlier observers, "many of them possessed of good telescopes," strangely enough did not record them or "recorded few of the domes." Wilkins often refers to this nickname "bowler domes" for to him "they look for all the world like [English] bowler hats." (*Our Moon*.)

Jessup cites many examples of such domelike structures suddenly appearing on the Moon. Here are a few of the more startling ones:

September 1889—Professor Thury of Geneva observed the sudden appearance in the crater Plinius of a "circular, chalk-like spot in the center." Jessup concludes: "Clearly this was a UFO. What else could something of a temporary nature with a little spot in the center be?" he asks.

To Jessup such bright, circular white spots, which are described as domes, are UFOs.

The British astronomer Birt in 1879 noted "bright spots" appearing in crater Hyginus. Astronomer N. E. Green also observed dramatic changes in the crater Hyginus. In *Astronomical Register* (Vol. XVII, p. 144) he noted that the crater changed its brightness from night to night and that "evidently no crater or hollow" seemed to exist there any longer "but seems rather a spot of color instead of a crater, but this cannot be the case because it is lost when the sun rises."

Other astronomers have noted rapid changes here "showing markings which changed their appearances completely in twenty minutes." This is too rapid a change to ascribe to anything else but intelligent control, concludes the astronomer Jessup.

Astronomer Elger described "a bright round spot" suddenly appearing in the crater Fracastorius. In a different light it resembled a low, circular hill or "round-topped tableland." It was domelike—one of the many domes which suddenly appeared on the Moon. But this one suddenly took on a remarkable appearance, different from some of the others. Notes Jessup: "It seemed to be surrounded by a peculiar glow quite different from the lights of other spots on the floor of Fracastorius, and in the center of the glow I could just distinguish a delicate crater of the most minute type, which would certainly not have been visible had not the definition been exceptionally good."

Similarly, strange "bright round spots" have appeared all over the Moon. Many bright round spots appeared in areas of strange activity like Plato or the Sea of Crisis. Commenting on the strange "very bright round spots" suddenly appearing in the largest of the craterlets on the floor of

Plato, where so many bright lights have been reported, the English astronomer Wilkins notes: "Why an object which usually requires some looking for should have suddenly become a large bright spot is a mystery. It looked as though a little craterlet was filled with something which strongly reflected sunlight. Whatever it was, it completely altered the usual appearance, turning a well-defined craterlet into a bright spot."

Jessup says of this remarkable change in the Plato Crater: "Wilkins seems to have failed to note the parallelism here to Linne and other 'bowler hat' phenomena which he himself has mentioned."

Wilkins, of course, unlike Jessup, does not state that these are definitely evidence of alien intelligence on the Moon. In the book *Our Moon* he freely admits his puzzlement: "We cannot subscribe to this idea because without air to breathe it is exceedingly difficult to contemplate the existence of selenites let alone to speculate as to their possible activities, industrial or otherwise. It is equally difficult to explain these things on natural grounds."

Jessup, who obviously has a high regard for this great English astronomer's judgment and refers to his books as "one of the most modern of lunar treatises and most generous to lunar activity," does not quite agree with Wilkins, but notes that these strange domelike structures began to appear over a hundred years ago although in sparse numbers.

Wilkins does point out that the "foremost seleneographer" of the nineteenth century, Nasmyth, claims that none existed apart from one sighted north of Birt Crater near the Straight Wall. Wilkins, however, notes that "today they are known [there] in considerable quantities." He claims he and Patrick Moore alone have discovered nearly a hundred. Wilkins claims that domes increased in number every year during the fifties. Jessup writing in the fifties, claimed: *"Their number doubles just about every twenty years."* (*The Expanding Case for the UFO.*)

Jessup agrees with Wilkins that these domes are indeed mysterious. He agrees essentially with the great British astronomer that "in spite of his own extensive study and review of all other reports, he cannot explain the domes."

Jessup does point out, however, that there were only "two in 1865." By his own time, in the 1950s, there were over 200. Some estimate today there are nearly 1000!

One of the earliest and most impressive sudden appearances of a dome on the Moon took place in the well-known crater Linne, which had been mapped by a great number of astronomers and then suddenly disappeared. In the mid-1860s dramatic changes began to take place in Linne. In the 1840s Johann Schroeter had mapped a 6-mile crater named Linne. It had a depth of about 1200 feet.

This German astronomer made hundreds of the maps of the Moon over many years, yet in the instance of Linne his observations noted that this crater had, strangely, disappeared. Then, in November 1866, changes began to occur in Linne with puzzling rapidity. On November 13, 1866, the astronomer Webb described "an ill-defined whiteness" he saw there. Talmadge claimed he saw "a circular dark cloud." The astronomer Denning claimed that on January 8 he saw a small hill rise out of Linne, appearing to be a "very brilliant point of Light." Later Denning concluded that a "hill" existed here. He even described "a white cloud which had replaced Linne."

What do astronomers make of all this? Jessup claims if you combine all the descriptions you come up with artificial structures of some kind. He believes they could be UFOs.

In *The Expanding Case for the UFO* Jessup gives page after page of the map changes and data on how Linne changed in size and structure, even on the dome itself that "covered" Linne.

He says: "So the 'thing' covering Linne was fluctuating in size as well as brightness." He asks: "If this was a real cloud [as some described it] what prevented it from dissipating into the vacuous atmosphere of the moon?" He notes that it was too tenuous to cast a shadow, yet opaque enough "to obscure the surface below it." How to account for all this?

Jessup quotes the astronomer Wilkins: "Today Linne is the reverse of a crater, being in fact a hill or dome with a minute pit on its summit."*

Jessup, who believes Linne is definitely an artificial alien

---

*As noted before, NASA claims that photos of this area show no dome-like structure but only a tiny, pitted area.

construction and maybe even a UFO, asks: "Was Linne the first 'colony'?" That is, the first of returning "Moonmen" in modern times. He insists that intelligences have been on the Moon from time to time at least since the early beginnings of human civilization.

Jessup makes a striking observation about this strange domelike structure which he claims is true for all such constructions. "There must be some significance in these unfocusable spots and nebulosities on the Moon. What prevents a telescope from receiving a sharp image when the surroundings are perfectly clear?"

Concludes the orthodox scientist who is able to shed his orthodox views: "This is the first report of this series of UFO operations—the disappearing craters, superimposed nebulosities (clouds, mists) and the rest. . . ."

Strange clouds of all kinds, shapes, and sizes appearing and moving around the surface of the Moon, some with lightinglike speed, have been recorded, as the NASA report examined in Chapter 2 noted. What does Jessup make of these?

This bold astronomer tackles the problem head on: "How solve the enigma of small clouds associated with minute craters on the surface of an orb which has practically no atmosphere? Something must be generating and dispersing these clouds that appear and disappear in a manner not consistent with a highly rarified atmosphere."

With the present knowledge we have garnered from our recent studies of the Moon, we now know for sure that clouds are absolutely impossible on this airless world.

Yet "clouds" have been seen. What are they?

Take a typical sighting reported in great detail by the English astronomer Birt. He notes that he saw "a white-cloud-like spot" appear suddenly near Picard. (*Astronomical* Register, 1864, Vol. II, p. 295.)

Birt comments on this "cloud": "In the course of my observations, as I observed the locality under oblique illumination, the white cloud spot became invisible or did not exist; which, I can not say. But its want of definiteness and its similarity in appearance to a cloud led me to hesitate before expressing an opinion as to what it really appeared to be. Further observation brought to light a small pit-like depression in its neighborhood with which the larger cloud-like marking appeared to be connected. The pit-like depression is of a beautiful whiteness and shows up when the cloud is not visible."

Jessup identifies these "clouds" with the strange domes of the Moon. He says: "This looks very much like the lair of some sort of UFO." Both the "clouds" and the "domes," he is convinced, are some kind of UFO.

For hadn't a number of astronomers described the strange dome that appeared at Linne as a "cloud"? In 1866 the astronomer Schmidt announced that the well-known Linne Crater, which had been mapped by a great number of astronomers before it suddenly disappeared, had turned "into a whitish cloud." Significantly he added, it "appears to be moving around."

This surely sounds like a UFO. And Jessup concludes that it is exactly that!

Unlike other astronomers, however, Wilkins does not pass them off as purely natural phenomena or deny that they even exist.

In this regard the great English astronomer remarks somewhat acidly:

> Of course, the cynics will declare that these peculiarities have no real existence, that it is all a case of observers being mistaken. However, it is a remarkable thing that the people who refuse to believe these things are the very people who have the least experience in observing the Moon. Some of them have never seen the Moon through a telescope. Perhaps this is wise, for if they did they might be converted.
>
> It is like the contemporary of Galileo who declared, having proved to satisfaction that satellites or moons of the Planet Jupiter did not exist, and then would not look through Galileo's telescope in case he should see them. He died shortly afterwards, which prompted Galileo to remark that it was to be hoped that he saw them while on his way to heaven! (*Our Moon.*)

Wilkins does not make the most direct conclusion, which Jessup does, about all these domes, clouds, and other strange "events" taking place on the Moon. For Jessup concludes that they are evidence of alien beings on the satellite of Earth. Interestingly, Wilkins does not reject this possibility. In fact, he and colleague and coauthor, Patrick Moore in their book *The Moon* confess: "It is not impossible that on the Moon there may exist, or have once existed, some form of life peculiar to the Moon and totally unlike anything known on Earth."

But Jessup is bolder. Studying the evidence of lights, clouds, and domes on the Moon, he hesitates not a single moment before proclaiming them to be the evidence of alien beings. He notes the close association of the inexplicable disappearing craters with bright spots, domes, lights, and "clouds." "Note the frequency which spots, lights and clouds appear in groups of one to eight or nine—on the Moon." (*The Expanding Case for the UFO.*)

To Jessup, the crack astronomer, there can be no other conclusion: They are UFOs and evidence of alien intelligence on and probably coming from inside this strange world of the Moon.

STRANGE DISAPPEARANCES ON THE MOON

Some astronomers may argue that the appearance of a cloud of dust is merely a meteor hitting and "kicking up" Moon particles on the surface. Until recently others insisted that the sudden appearance of a dome or mist was due to volcanic activity, a theory eliminated by scientific evidence that proves our Moon has been a dead world for eons—volcanically dead, that is.

Then how to explain the disappearance of a structure on the Moon? And these have occurred. On the edge of the Sea of Cold, not far from the strange crater which has been the center of so much activity on the Moon, Plato, is a large, almost perfect square, described by astronomer Madler in detail. Madler says that within this rampartslike square structure of four walls "was an almost perfect cross formed by white ridges."

Yet today it is gone. Wilkins, using the greatest telescope in Europe, reports confidently that one side of the square no longer exists and that the cross has disappeared! The area has been searched and searched with telescopes ten times the diameter of the one used by Madler. Says Jessup:

"Something, then moved away a huge wall and a great cross. What could that be—alien intelligence?" (*The Expanding Case for the UFO.*)

Jessup then quotes the British astronomer Wilkins, whose puzzlement we have already noted, as candidly confessing that we cannot completely rule out the possibility that these changes on the Moon's surface, this seeming "intelligent activity" indicated by "events" and "happenings" on the Moon's surface, could possibly indicate the existence of intelligent beings on the Moon. Although Wilkins does not take the final step and conclude that intelligent beings exist on the Moon, he does admit that these "events" cannot be explained away "on natural grounds."

Jessup, who has sincere admiration for Wilkins, observes: "This is an active mind striving against the fetters of dogmatized science. This is as far as we may ask an astronomer to go in advocating intelligence on the Moon, or anywhere else in space. More would cut him off from his profession. Let this man of authority state the observational facts. I, already an outcast, will cheerfully take the rap by stating bluntly what the facts imply—*UFO activity!*" (*The Expanding Case for the UFO.* Emphasis added.)

Most of these sightings took place several decades ago; some centuries ago. Is there any evidence that there are intelligent beings on the Moon today?

Before Jessup died in the late fifties he also authored a number of UFO annuals. In his *UFO Annual for 1955* he reports that two amateur astronomers saw a strange happening on the Moon on the night of July 5, 1955. One observer saw a strange light "hover for three minutes and then move off the Moon."

Jessup concludes: "That's controlled motion and control means intelligence!"

One of these astronomers reported that the spot of light appeared at the exact center of the Moon's disk; the other said it had moved to the Moon's edge. That established two base lines, one on Earth, one on the Moon, notes Jessup. With these two lines and the known distance from the Earth to the Moon it was possible to locate that UFO's position in space. It was 95,000 miles out.

This is perhaps the most amazing sighting and evidence Jessup comes up with to indicate UFOs in the vicinity of the Moon.

In our book *Our Mysterious Spaceship Moon*, we noted the many sightings of our Apollo astronauts. But if we are to believe an account related by a former Russian space scientist who defected from the Soviet Union to France in 1969, the Soviets had actually landed men on the Moon. The story is interesting because the claim is made that while there Soviet astronauts had a strange encounter with "alien intelligences."

In an interview carried in *Beyond Magazine* (February 1969) Professor Lev Mohilyn, who claims he participated in the Soviet space program before his defection relates the account of that unknown Soviet mission. On June 5, 1968, according to his story, a manned Russian rocket was launched from a base in the Ural Mountains. Its destination: the Moon. It carried two cosmonauts (Ilya and Evgeny), who successfully made it to the Moon in three days. After accomplishing a soft landing they left their craft to carry out some explorations on the lunar surface. However, while out there, according to Professor Mohilyn, a "mechanical monster" attacked Evgeny and killed him. Ilya rushed back to the spaceship and upon orders blasted off the Moon's surface. She (Ilya was a lady cosmonaut) eventually safely completed the return trip to Earth.

Is there anything to back up Mohilyn's incredible story? No one knows. It seems to be utterly preposterous on the surface, and while it is true that the Soviets never (at least to Western world's knowledge) ever claimed to have achieved a manned landing on the Moon, unquestionably they had the power and capability to attempt such a landing. But it is doubtful that they could have accomplished a successful landing at that early a time. Their journey purportedly (if we are to believe Professor Mohilyn's account) took place a full year before Neil Armstrong and Buzz Aldrin set foot on the Moon.

RADIO SIGNALS FROM INSIDE THE MOON?

Another unauthenticated account claims that radio signals have been received from *inside* the Moon.

Back in the late thirties, radio engineer Grote Reber allegedly was trying out a radical new type of radio telescope experimenting for the Bell Telephone Company in New Jersey. The purpose of the experiment he claimed was to collect and "focus radio waves from space." Riley Crabb of the Borderland Sciences Research Foundation who relates this incredible account, notes that although "orthodox astronomy says the moon is a dead place," according to Bell Telephone engineer, Karl Jansky "something is happening there; for Reber turned his radio telescope on the Moon and did receive signals from the Moon. This was years ago, before such information was classified top secret. Jansky said the radio signals originated from *inside* the Moon, not on the surface. This would indicate that the interior of the moon is inhabited," concludes Crabb. (*Meeting on the Moon*, Borderland Sciences Research Foundation, 1964.)

Though this story cannot be verified, it is interesting to note that other reports of radio signals emanating from the Moon have been verified. In our book *Our Mysterious Spaceship Moon* we noted that in 1927, 1928, and 1934 "mystifying radio signals were detected in the vicinity of our companion world, the Moon, by various radio researchers."

In 1935 two scientists named Van Der Pol and Stormer detected radio signals on and around our Moon. Even radio and electrical geniuses Marconi and Tesla turned in some strange reports.

Perhaps the most recent such radio signals to be reported to our knowledge occurred in 1956 and were picked up by Ohio University and other observatories around the world. At the time Ohio University researchers claimed to have received a "codelike chatter from the Moon." (*Saga*, April 1974.)

Whether the radio evidence is valid or not, there appears to be no doubt that the cumulative tons of evidence of thousands of strange lights and moving objects on the Moon and the sightings of our Apollo astronauts, as well as various other strange things happening on the Moon, do indicate that without a doubt not only is our Moon a spaceship but it still has intelligent beings aboard!

ARE ALIENS TAMPERING WITH OUR SPACE EQUIPMENT?

There have been some strange things happening to our space equipment on the Moon which might indicate that aliens are still there.

In April 1976 the *New York Times* News service carried an intriguing article about a startling development that took place on the Moon. "Mysterious Force Pesters Apollo Station" (*Detroit Free Press*, April 22, 1976.)

The seismometer station which had been set upon the Moon by Apollo 14 astronauts (one of five deployed there) had been operating without a hitch since February 1971. Then suddenly it went dead in March 1975. Naturally NASA officials were not surprised when it stopped working. When the radio receiver failed in March 1975 and the transmitter quit working on January 18, 1976, NASA scientists were anything but disturbed. The Apollo Lunar Scientific Experiments Package (ALSEPS) scientists had thought it would last only about a year. Strangely, the atomic power supplying the ALSEPS lasted much longer than scientists anticipated.

The mystery, however, began when about a month later (February 19, 1976) the Apollo 14 ALSEPS equipment suddenly began working again. Furthermore, the mystery deepened for not only did the radio transmitter and receiver begin operating perfectly once more, but one of the ALSEPS experiments that had never operated in the intense heat of lunar daytime now began "operating flawlessly night and day."

Then, according to the *Times* news service, one month later "the mysterious force shut the whole station down again." Charles Redwood, NASA spokesman out of Houston, candidly admitted: "It's a bit of a mystery. We have a number of people trying to figure what's happening up there but we haven't got an answer yet."

Are we to conclude from this "mystery force" pestering one of our Apollo stations that someone is on the Moon? There is undoubtedly another explanation to this Moon mystery. But other far more substantial and impressive evidence exists to indicate that our spaceship Moon is presently occupied.

THE APOLLO 16 MYSTERY

A similar incident happened to the Apollo 16 spacecraft which a few sensationalistic writers seized as evidence that intelligent "Moon beings" were trying to play havoc with our space probes.

Our Apollo 16 lunar spacecraft was suddenly thrown off course. Richard Lewis a reliable science reporter, says: "No one knew what happened to throw the inertial guidance system into a lock. It was probably one of those mysterious, electrical transients called a 'glitch' that no one really understands but that create temporary malfunction in electronic equipment and then disappear without a trace." (*The Voyages of Apollo*.)

Modern People Press Special UFO Report (Spring 1976) includes another strange incident in an article intriguingly entitled "Unexplained Lunar Mysteries Point to Intelligent 'Moon Men.'" They note that a "mysterious electrical impulse" stopped our unmanned Ranger III spacecraft from taking television pictures after it landed on the Moon. There appeared to be no malfunction, but something countermanded an order radioed by the NASA Goldstone Tracking Station. Apparently no explanation was forthcoming for what happened and no one could "even guess from whence the impulse emanated."

Do such accounts indicate "intelligent Moon men"? Hardly. There probably are rational explanations for them which explain them on natural grounds. However, as we have seen time and again, there is overwhelming evidence to support the conclusion that a strange alien presence does exist on the Moon, as shown by the many strange, unexplainable encounters the Apollo astronauts experienced there. Almost every Apollo astronaut saw strange lights, unidentified flying objects, or heard strange radio noises or sounds. This plus the astronomic evidence of definite activity on the Moon in the past several decades as well as centuries, continuing to the present time, would lead any open-minded investigator to the conclusion that someone is up there on the Moon. Lights and glows, strange structures, flying objects that move and hover above the surface of the Moon seen both by astronauts and astronomers, cannot all be explained away as optical illusions or hallucinations. No, someone is occupying the Moon. And they may have a great deal to do with you and me—and our home, Earth.

. . . I think people are hungering for mystery, and the moon has been a place of mystery for man as long as the human race has existed.
—*Astronaut James Irwin*

FIFTEEN

THE MULTIPLYING MYSTERIES OF THE MOON

Our view of the Moon may be considered revolutionary but we must remember that our own view of our own planet Earth today would have been considered science fiction a mere two centuries ago.

Less than two hundred years ago the scientific world thought that our planet was but a few thousand years old. Then scientists like Charles Lyell (1797–1875), the father of modern geology, began to see that, contrary to common thought of the time, changes on Earth took place slowly over eons, rather than quickly; that indeed Earth was hundreds of millions if not billions of years old.

This doctrine of change in the "chemistry of time" (called uniformitarianism), says science writer Richard Lewis in his intriguing book *The Voyages of Apollo,* "was the key to understanding the Solar System. Without this idea it would have been a waste of time and money from a scientific point of view to have explored the Moon, or any other planet. *One simply could not have understood what one was looking at.*" (Emphasis added.)

This may be still true today in regard to the Moon. Scientists locked in orthodox methods cannot shake old ways of thinking.

Concludes Lewis: "*It may be the mysteries of the Moon still eluding science require for their solution a new conceptual departure as radical as uniformitarianism was in the eighteenth century.*" (Emphasis added.)

Urey told his fellow scientists just before our manned explorations of the Moon began: "Let your imaginations go . . ."

This does not mean to give free rein to the imagination without sticking to the facts and the evidence, but merely to allow for consideration other imaginative ideas or theories about the origin and nature of this mysterious Moon of ours.

Dr. S. Agrell of the University of Cambridge in England said the same thing: "It seems to me that lunar geology is something which is quite distinct." Agrell sounded the warning to his fellow scientists: "Clinging to earthly geology is like wearing a special kind of blinders, which do not allow the scientist to be imaginative enough." (*Science News,* January 10, 1970.)

Yet scientists today for the most part remain straitjacketed in the old orthodoxy—locked in old ways of looking at the Moon.

In 1975 when this author published our ground-breaking investigation of the bizarre Soviet artificial-Moon theory, we hoped ardently it would send "shock waves throughout the scientific world," as the publisher predicted. Copies of the book were sent to just about all the leading scientists quoted extensively in the work. Unfortunately, some, like the late great Dr. Paul Gast had passed away in the interim.

Yet very few of those scientists who received a copy of the book took the trouble to respond to this striking Soviet thesis. One well-known scientist and space expert returned the manila envelope (in which the book was sent) unopened. It was marked in his handwriting: "Rejected." Apparently he had seen my name on the return address, and although he knew me not, probably had heard I had quoted him in this shocking book and this was too much for his sensitive closed mind. He refused to even accept a complimentary copy of the work.

Most did accept it but never bothered to respond one way or the other. Probably most passed off the book as just the work of another crackpot—in this case "lunatic." The Moon couldn't possibly be a spaceship. They undoubtedly rejected this book despite the fact that the work is based almost entirely on factual evidence that both our American space scientists and the Soviet scientists themselves documented. All we did was cite the facts and apply it to the Soviet theory—which, by the way, we might point out is the brainchild of two *orthodox* researchers, not ours.

Only a few scientists bothered to respond; still fewer admitted that the Soviet spaceship theory could be correct. One scientist did however, go so far as to encourage me to continue my research work and publish the facts whatever they might be.

Realistically, however, the theory is just too bizarre to be given open-minded serious consideration. It is just simply too far out, too "science-fictionish" to be given even passing consideration by orthodox scientists.

One aging dean of lunar experts did acknowledge receiving and reading the book—at least in part. That is, he read it up to the statement that the "Moon was a spaceship." That turned him off immediately. Then he added:

"I have not read your entire book . . . [but] read through to the suggestion that the Moon was made by human efforts. This, of course, cannot be true. Man has a great deal of capacity for doing things, but making a moon is beyond his capacity. It just cannot be a correct explanation."

Although this aging scientist is one of the top men in his field, he did have the courage to freely admit that after all his and his colleagues study, the Moon remains a complete mystery to man. Still, he could not bring himself around to even considering an unorthodox theory.

Of course, the asinity of the entire episode is pointed up by the fact that he was so closed-minded he did not even read far enough to realize that in no way did either this author or the Soviet researchers who formulated the theory ever suggest the Moon was "made by human efforts." Unquestionably it was made by alien intelligences who are obviously well beyond the level of human technology. Either the good doctor misunderstood the entire thrust of the theory or with his straitjacketed orthodox mind he felt if the Moon, as the Soviet scientists insist, is "the creation of alien intelligence," this must mean man, for in his mind, as in many scientists' minds, the only intelligent being in the universe is man.

Undoubtedly much is left to be ferreted out. With the secrecy enshrouding our space program, however, this might be difficult to accomplish. For as we have pointed out, our government and its space agency have indeed been covering up the truth, hiding it from the public mind probably because of the fear of panic that such frightening news might bring.

It is left for the people themselves to bring it all out into the open. You can help. Maybe if enough people marshal themselves and mount enough pressure then we can crack this shroud of secrecy wide open.

After the publication of my first book on this subject, *Our Mysterious Spaceship Moon*, the head of a Midwest education association wrote and told me he found the work "thought provoking but very troubling."

He said: "Like many, I have suspected for quite some time that several of our government agencies have been 'keeping things from the public' regarding discoveries made during various of our space explorations. Your book only serves to confirm and deepen that suspicion. I wish that there were some way that citizens could press these governmental agencies for further information."

There is, of course, always the opportunity to write to your elected officials, whether congressmen or President. Many of the more highly respected UFO organizations (Dr. Hynek's Center for UFO Studies, APRO, NICAP, and MUFON) have over the years continued to criticize the government's cover-up policy in regard to the existence of UFOs.

In regard to the Moon there is another way. If sufficient evidence were collected of the more sensational kind in regard to the governmental space program cover-up, then perhaps the people would be aroused enough to get what we're supposed to have in the country, "a government of the people, by the people and *FOR* the people." This country was designed by our Founding Fathers to be free and will survive only with a government that is open and aboveboard in its dealings with the people. As citizens of this land we have a duty to find out what is going on, to learn the truth whatever it might be.

THE GREAT WORK OF LUNAR SCIENTISTS

We do not in the least intend to demean the outstanding work that our dedicated scientists have already done. Indeed, these quiet heroes devoted to learning the truth about the Moon have worked wholeheartedly for years, mostly with frustration but not completely without success.

As astronaut Eugene Cernan, the last man on the Moon, told a conference of space workers at Cape Kennedy in 1973: "Apollo 17 had the shoulders of giants on which we stood as we reached for the stars. . . . These were your shoulders and we thank you for them."

I sincerely wish to acknowledge that in the writing of this work (as well as my first book on this subject, *Our Mysterious Spaceship Moon*) we too have had the shoulders of giants that enabled us to divine the truth about our Moon. We thank these many giants of science whom we freely quoted and whose work we so frequently cited in these books. We thank them—for without them these works could not have been written.

In fact, we must humbly point out that we did not originate the artificial-Moon theory. Far greater minds than ours accomplished this feat. The author, in fact, is only an open-minded curious student of life who chanced upon it and who followed the facts wherever they led—in this case, it would seem, to an unbelievable truth.

In the truest sense, I am merely a compiler, an organizer, a synthesizer, as it were. All credit must go to those scientific giants—first the two Soviet researchers who originated the theory, and secondly to the army of international lunar experts whose hard work and dedication ferreted out the facts and the findings which filled in the skeletal outline of the Soviet spaceship theory. The artificial-Moon theory is really their brainchild. I am merely its mental midwife.

It is with this in mind that I appeal to my readers, inviting them to communicate to me any direct or indirect knowledge of any such cover-ups. As we have seen, the Moon is swirling with mysteries. Who knows but focusing the public attention on any one of these revelations might be enough to blow the lid on the entire secret closet of governmental space cover-ups.

The evidence we have cited in this book proves that we have on our hands today another Watergate—a cosmic Watergate. UFOs are swirling around our planet and official governmental agencies have tried desperately to pass them off as mistaken objects or the product of public hallucination. There appears to be sufficient evidence to conclude that the Moon is the base for these UFOs coming to our Earth. There is not much doubt about it—we are under observation! For what purpose we can only guess. We believe, however, that the people should be attuned to truth so that if something shocking happens in this regard tomorrow, people will be ready for it. Otherwise, if a sudden, shocking revelation does come, that might produce panic.

AN APPEAL TO READERS

Perhaps one of the most telling ways we can crack the curtain of secrecy is to bring to public attention any shocking examples of cover-up of a startling nature. After my first book on the Moon came into print a number of interested readers took the trouble to write me, expressing their concern. Some of them imparted tidbits of information that proved useful.

An editor of a prominent national magazine in New York asked me bluntly if I had ever run across any accounts of a huge space platform which our lunar astronauts were supposed to have sighted on the far side of the Moon. He had heard rumors that one many miles across had been seen but was covered up by the space agency. At the time I told him frankly I had not. But the story piqued my interest and led me to do a little digging. Now we know the astronauts did see such a structure (see Chapter 3)—at least according to some reports.

Another magazine editor told me that several readers had written to him claiming that one of the astronauts (on television) had "picked up what looked like a glass bottle." The astronaut was supposed to have exclaimed as he did so: "My God, I don't believe it, look at this . . ."

However, those who communicated this story to the editor claimed that the television screen suddenly "went blank and into some news . . ."

Is there anything to this story? Or this one: A high school student once enthusiastically related to me the fact that he had a complete collection of tapes recorded from television commentary of the various Apollo explorations of the Moon. He said that one major network was describing a drilling operation on the Moon. The astronauts, although armed with drills that could go through just about anything, were having extreme difficulty penetrating even a few inches of lunar crust. Only after great effort did they drill into the maria surface and with even greater exertion extract this drill. It took prodigious efforts of both astronauts tugging at the drill, but when they finally succeeded in pulling the drill out, this young man insisted, the television commentator exclaimed: "Look at that drill, it's encrusted with metal shavings."

I anxiously awaited the tapes, but finally he told me that they were not available. He claimed his kid brother had mistakenly taped over the invaluable Apollo legacy to get some of his favorite rock hits on tape.

Is there anything to the story? We don't know. But if it can be proven that this alleged account is true, this one item could break the whole Moon cover-up wide open. This is the kind of evidence that also could crack the mystery of the Moon wide open.

If any reader has knowledge of this or any other mystifying happening in regard to the Moon or the space program, I will be more than glad to hear from you and correspond with you about your discovery.

I would like to steer away from anything of a religious nature. One goodhearted soul offered to give me the spiritual secrets of the universe, including the Moon, that she claimed she got directly from God.

Another related this strange Biblical coincidence: The initial letters of the names of the first three men on Earth (Adam, Abel, and Cain) are the same as those of the astronauts who made the first landing trip to the Moon (Armstrong, Aldrin and Collins, although the latter never walked on the Moon but stayed in orbit).

HEAVENLY MUSIC?

Along those same lines but more to my liking was the story related to me that our astronauts while on the way to the Moon actually heard strange music over their closed radio communication system. This party maintains that the cabin of the space capsule was filled with beautiful haunting music that seemed to come from out of nowhere. Later the "heavenly" music was identified. It was an old Earth-produced piece: "Where Angels Fear to Tread."

VOICES ON THE MOON?

A lady from California related that a well-known French author Robert Charroux in his book *Our Mysterious Past* points out that "mysterious words [were] heard on the moon."

Charroux claims that a famous French science journalist reported that while Al Worden was walking on the Moon (Apollo 15) mysterious words mystifyingly interrupted the CAPCOM (Mission Control) communications network. Apparently, according to Charroux, it was written up widely in the French press, although I have never seen any reference to this incident in the American press. The French broadcasting network ORTF also carried stories about this strange happening, Charroux maintains.

Charroux also claims that the French weekly *Le Meilleur* (edition 33, page 1) published an article seven columns long, headed "Why Has No One Spoken of the Mysterious Message Heard on the Moon?" They tantalized their readers with this subtitle: "Twenty Untranslatable Words Which Really Sowed the Seeds of Panic?"

"Perhaps this proves that other men exist—something

NASA wished to hide," concludes the French journal.

What were the words the astronauts were supposed to have heard? On August 3, 1971, at exactly 11:00 A.M. without explanation the normal space radio communications faded strangely away and contact with Houston Mission Control was lost.

Worden then noticed strange sounds coming through his receiver—a weird breathing sound which gradually changed into a kind of whistle. After "stifled murmurs," vague sounds resembling words came through—then a string of words—"like a sentence constantly repeated on one note."

It is claimed that the transmissions were recorded on the lunar module's tape recorder and Worden transmitted them to NASA authorities.

The entire affair was purportedly suppressed and a news blackout apparently was effective in this country as well as throughout the rest of the world. Charroux insists that the story in *Le Meilleur* is accurate and that "a conspiracy forbade the divulgence of the 'moon sentence.'" He then divulges the "forbidden sentence"—which no one yet has been able to translate.

His text is not too clear on this point but apparently Charroux who claims he himself heard the message given over French radio included two Hebrew words.

"We thought we had remembered two words from the text: 'Lamma'—and 'rabbi.'" Charroux points out that a similar word, "lama" was used by Jesus while he lay dying on the Cross (ELI, ELI, LAMA SABACHANI—My God, My God, why hast Thou forsaken me?). Apparently, however, there were some misgivings that this word was actually heard. For he quickly adds: "It is possible we misheard the first word of the sentence which could be 'mara' and not 'lama.'"

So the message as Charroux *remembers* it is:

"MARA [OR LAMA] RABBI ALLARDI DINI ENDAVOUR ESA COUNS ALIM."

Although Charroux claims some of the words might be of Hebrew origin he admits all the others seem to be of "uncertain origin."

In fact, from his entire account we can conclude everything is "uncertain." Not only uncertain in content and unknown in meaning but uncertain in authenticity and unknown in import. No authentication for this event has ever, to our knowledge, been forthcoming. This is not to say it did not happen. It might have. If anyone has any further information on this we would be happy to hear from you.

Charroux concludes with his usual candor: "Perhaps some philosophers will find the key to the puzzle."

Perhaps our readers will find the answer. Or similar material—hopefully better authenticated—which they will relay to us that we may work a little harder in cracking open the mystery of our Moon.

For it appears that it is not going to be done by orthodox scientists. Nor will the space agency change its policies without our help. They have probably garnered sufficient evidence, but as we suspect, they are keeping it under wraps.

It will take the efforts of people like you—open-minded searchers for the truth—to find the truth whatever it might be.

The fact is that sufficient public clamor or some dramatic discovery might not only break open this cosmic Watergate, exposing the greatest cover-up of all time, but might lead to a reopening of our investigation of the Moon again and possibly even future lunar probes.

Shortly after *Our Mysterious Spaceship Moon* was published the author was interviewed on a radio station together with an astronomer of a well-known Midwestern observatory. He agreed, surprisingly, that the Soviet theory —which he did not accept for lack of sufficient evidence— could possibly be correct, that the Moon indeed could be a spaceship. He admitted freely that, strangely enough, the dozen mysteries delineated by the book did disappear in light of this bizarre theory. He told me after the radio interview: "Wouldn't it be remarkable if your book arouses the American public enough to cause the Moon-exploration missions to be resumed?"

"That would be great," I told him. But realistically, of course, this is not going to happen. Undoubtedly man will go back to the Moon someday and when he does there are a number of fascinating and probably eye opening areas that would be worth visiting. Or should we say "revisiting," since with the space program's secrecy we cannot be sure whether or not our astronauts and unmanned space probes have not already checked them out.

In *Our Mysterious Spaceship Moon* we listed a dozen "mystery spots" on the Moon worth visiting. One of them was the strange 12-mile bridge sighted over the Sea of Crisis back in the fifties (see Chapter 2). Dr. Farouk El Baz was quoted in a national magazine as stating that our space agency did carry out secret investigations of the Moon, and this included this strange artificial-looking bridge on the Sea of Crisis.* In similar fashion we went through other intriguing "mystery spots." Here are a half dozen more which might produce discoveries that could definitely prove the Moon is a spaceship.

PYRAMIDS ON THE SEA OF TRANQUILITY?

First and foremost, the strange pyramidal, obelisklike structures that appear to be artificial and which Soviet space engineer Alexander Abramov claims are positioned exactly like the major pyramids around Giza near Cairo, Egypt. These have probably been closely examined by NASA, since our astronauts went to the very same Sea of Tranquility on our first trip to the Moon, only the results have never been released and probably never will be unless enough public pressure can be applied on our government.

A SPACE ODYSSEY MONOLITH?

Secondly, the site of the Soviet-discovered monolith, that mysterious rectangular block of stone that appears strangely enough to be like the monolith of *2001: A Space Odyssey,* Arthur Clarke's famous book and movie. (See Chapter 17.)

Thirdly, *the inside of "Ranger" Crater.* The controversial "Ranger 7 Crater" would be a good place to concentrate

---

*NASA claims that photos of the area where the purported bridge was spotted showed nothing unusual.

our Moon look, this time without the veil of secrecy covering up our findings.

The first phase of our Moon exploration the Ranger series returned thousands of photos of the Moon's surface —close-up pictures that revealed a great deal about the surface of this strange satellite. One of these taken by Ranger 7 just before it smashed into the Moon has produced a storm of controversy. The photo was taken about three miles on the last leg of its crash dive and appears to show some objects inside a crater. NASA officials claim they are just "a cluster of rocks." Other investigators are not so sure.

No one knows for sure but some investigators like Riley Crabb of the Borderland Science Research Organization claims that their circular symmetry indicates that whatever they are, certainly they are "intelligently constructed."

Crabb believes that this conclusion is "confirmed by the sharp, straight black shadow cast" by one of two "brilliant white shafts" inside the crater. He points out that this August 10, 1964, photo published in *Missiles and Rockets* magazine (p. 22), shows "clearly outlined in between the bases of the two shafts a perfect circle, perhaps 40 or 50 feet across. The hole itself is pitch black, as though it led into the interior of the moon; but the edges are bright, like the edges of a gigantic bubble or lens."

Crabb goes on to explain: "As we looked at the *Missiles and Rockets* reproduction of the Ranger photo we became aware of two more perfect black holes in the crater, one to the left of the left-hand shaft, and one in the top of the little cone to the south of the shafts . . ." He concludes: "You can be sure that military and civilian photo analysis have devoted hundreds of hours to this particular 'rock' formation, but in view of the continuing 'silence policy' on Flying Saucers such technical analyses will be classified secret for years to come." (*Journal of Borderland Research*, November–December 1964.)

Some see in this photo UFOs sitting inside this mysterious crater. One such investigator is George Leonard, an amateur astronomer who claims that through an ex-NASA scientist friend of his he got to look at a composite photo of several Ranger 7 pictures, that proves this is precisely what they are. This photo, says Leonard, put together by NASA's Jet Propulsion Lab in California, shows a large object in the crater with a "dull metallic finish." The object is "smoothly rounded, symmetrical, and has what appears to be a turret-shaped protuberance, which is also remarkable for its perfection."

Asks Leonard, with an obvious eye on UFOs: "What looks like that?"

He also is convinced that on its gleaming back he can discern a strange marking that looks something like an English letter Y with a line beneath it. He says a study of the world's alphabets indicates that it is very much like the ancient Semitic Z found on the famous ancient Moabite Stone.

If there is any validity to Leonard's conclusion its implications are staggering. This remarkable investigator is convinced that whoever is on the Moon was intrinsically involved in mankind's past.

"They've watched us develop since at least the Bronze Age. They've had a catbird seat on all our wars and pettinesses. They've architected and built big things here and left signs all over our Earth." So Leonard concludes.

Without hesitation Leonard in his intriguing book *Somebody Else Is On the Moon* speculates: "The Moon is a logical seat for all the UFOs skipping around the fringes of our cultures since the dawn of time." And in Ranger 7 photos he believes he has found evidence of such lunar UFOs, or at least of constructions on the Moon.

Surely here is a mystery spot worth taking another close look at!

THE STRANGE TUNNEL ON THE MOON

Another oddity of the Moon—and there are many—is a strange tunnel about 20 miles long which is lined with walls of glass!

Dr. H. H. Nininger, director of the American Meteorite Museum in Winslow, Arizona, announced this discovery back in 1952. He claims that through a good telescope not only can the tunnel be seen but the entrance and the exits of that tunnel are clearly discernible.

Located on the western part of the Moon in the Sea of Fecundity are two unusual craters Messier and W. H. Pickering. These two strange craters are very close together but they differ greatly from other lunar craters in that the rim or lip of each crater is "noticeably extended in the same direction." (*Science Digest,* November 1952.)

Dr. Nininger points out that the tunnel starts here with these two weird-looking formations, one the entrance and the other the exit. They are on the opposite sides of a towering mountain ridge, which is several thousand feet high and 15 to 20 miles wide.

The shape of the respective holes or entrances suggests that the same force has created a tunnel through this mountain.

Is this an artificial construction on the Moon? Dr. Nininger suggests that it was created by a "meteorite." This magical meteor "moving 20 to 30 miles per second would vaporize the powdery dust on contact," melting instantaneously and cooling quickly, thus forming a glazed tunnel.

It might seem impossible that any meteor or meteorite would do this which most scientists claim would explode on contact since even at the slow speed Nininger assigns to this slow-moving meteor it would still be travelling at 72,000 miles per hour! How could it then perform this miraculous construction job?

Another serious problem is that it is difficult to comprehend how it could travel horizontally to perform its tunneling job. But Nininger believes that a large meteorite could sweep low enough, skimming the Moon's surface, tunneling its way through solid rock mountain, dropping from the skies hitting and then ricocheting through this hard submantle, burning its way through, and leaving in its miraculous wake enormous holes that now "mark its entrance and exit." (*Science Digest,* November 1952.)

This unnatural explanation does not convince us that this is a natural construction. It sounds like a science fiction account if there ever was one. We are convinced that if and when we return to the Moon, as man someday in the future most assuredly shall, this mysterious tunnel would certainly be worth investigating. Who knows what it may lead us into? Perhaps final proof that indeed an alien world does exist under this natural rock.

THE MYSTERIOUS OPENING OR HOLE IN THE MOON

If we really wanted to get at the heart of the matter, there are the strange "plugholes," huge round openings which Dr. H. P. Wilkins of the British Astronomical Association is convinced may connect with extensive inner hollows of the Moon's cavernous interior. Are such connecting holes the passageways for aliens coming into and out of their hollow Spaceship Moon?

Wilkins claims that he discovered one such huge hole using one of the most powerful telescopes in Europe. He dubs it the "Washbowl," since it looks like one. This huge round opening into the Moon is located in Crater Cassini A and is over two football fields across.

In his intriguing book *Our Moon,* Dr. Wilkins gives us this remarkable description of this "hole" in the Moon: *"Its inside is as smooth as glass with a deep pit, or plughole about 200 yards across, at the center."* (Emphasis added.)

An opening like the mouth of a bottle certainly gives the appearance of being constructed. What startling discovery could be made here? Could this be one of the openings to the inner world of this Spaceship Moon?

THE PUZZLING STRAIGHT WALL

Earlier we mentioned another feature of the Moon which many astronomers in past centuries just assumed was an artificial construction—a straight wall, a ridge over sixty miles long which was so strangely straight it was nicknamed "The Railway." The Soviet scientists who formulated the spaceship theory of the Moon speculated that it might have been caused when the inner metal shell was ruptured. Thus, a huge armor plate inside the Moon, bending under the impact perhaps of a huge celestial torpedo, might have raised this straight, unnatural fault line of rock, pushing it outward.

The edges do present a steep rock cliff which rises from the surface of the Moon in a steady climb for over 1200 feet at about a sharp 45-degree angle. The area around it shows the ghost of a huge crater that may have crashed here eons ago. Intriguingly, all around are strange large, whitish domes.

Speculating on a mystery spot such as this can cause the imagination to soar. The author wrote a science-fiction novel (no, not this one!) depicting what the Moon would be like just after the turn of the twenty-first century, at a time when man had built three cities on it. Near this Straight Wall that the Soviet scientists think may have been caused by pushing up of the Moon's metallic armor plates, resulting in this unnatural, artificial-looking cliff of rock, astronauts land for a firsthand investigation. During blasting at the base of this strange Straight Wall structure, rocks fell away, revealing a metallic-beam construction underneath!

Soon thereafter, two of the astronauts fall through one of the openings here into the inner cavernous hollows of the Moon. A rescue mission discovers the shocking inside-out world of Spaceship Moon—its alien-looking cities and amazing centers of unimaginable construction. For these astronauts entering Spaceship Moon was like taking a journey into the future. I depicted this alien inside-out world of the inner Spaceship Moon as much like the Spaceship moon conceived by scientist-engineer Dandridge Cole in his book *Islands in Space*.

Man shall probably never get the opportunity to journey to this intriguing inner alien world of the Moon. But on the back side of the Moon, strangely enough, nearly directly opposite the Straight Wall is a huge crack 150 miles long and at places more than 5 miles wide!

If the Moon does have that inner metallic rock shell beneath its outer crust this vast opening could be very revealing. The position of this crack intrigues us into this speculation: could the gargantuan crack be related in formation to the Straight Wall? But this question is theoretical. The practical significance of the great crack in the

Moon is that it might be a great crack in the veil of rock enshrouding the hull of our Spaceship Moon.

Some of us believe the contours of an amazingly complex intelligent life beyond the Earth can already be discerned.
—Dr. Jacques Vallee

SIXTEEN
IS NASA HIDING THE REAL MOON FROM YOU?

The question has been asked time and again by the average taxpaying American citizen: Why did man go to the Moon? If you talk to scientists they will give you all kinds of different reasons. One of the most unusual answers ever to come from the scientific world was elicited from Dr. Marcus Langseth, a NASA geologist who hails from Columbia University's Lamont-Doherty Geological Observatory. Dr. Langseth is convinced that "going to the moon is an impulse ingrained in the natural character, as though Americans were astronautical lemmings. This vague feeling that the moon is pulling more toward it than just the tides is at the root of a great deal of NASA thinking," says this lunar expert. (Henry Cooper's *Apollo on the Moon*.)

You the reader, now having waded this far into the book, undoubtedly sense there were many more vital reasons that forced man into going to the Moon than just this fatalistic feeling that pervaded NASA scientists in Dr. Langseth's view.

The reader who has stayed with us this far and fought his way through the weighty, impressive evidence that indicates clearly that the Moon is a spaceship has been treated to some shocking revelations of our space program's discoveries about this strange world in our skies. Discoveries that you undoubtedly found fascinating and perhaps to some extent frightening. Moreover, revelations that you probably found disturbing. For the question of questions that undoubtedly arises again and again in your mind, if findings as staggering as those revealed in this book were made, if discoveries as shocking as these were made by our scientists, how come our space agency did not release this information to the media? Why were such fantastic discoveries with such staggering implications for man and his world buried or at the very least covered up?

Without a doubt some were. At the very outset of this book we showed how our space agency, contrary to its avowed public policy of openness, has, in fact, followed the path of secrecy and cover-up in many phases of its operations.

We showed incontrovertible evidence that NASA is hiding the fact that UFOs were seen by astronauts. That strange Unidentified Flying Objects were encountered by our astronauts in outer space can be proven by anyone who open-mindedly reviews the logs of the astronaut-mission control conversations.

Furthermore, if the tapes are not enough, we have the evidence and testimony of Dr. James Harder of the University of California, who maintains that after making this tape discovery himself he confronted NASA officials with this evidence and they privately admitted to him that our astronauts did in fact see UFOs on a number of Moon flights.

They also admitted that the government space agency covered up these sightings because of, as they put it, "fear of public panic."

The question now arises, could not our space agency and other governmental agencies be hiding the truth from the people about the Moon—again for "fear of public panic"?

Is it possible that the truth about our Moon's actually being a spaceship was surmised by both the U.S. and the U.S.S.R.—that they suspected the truth about the Moon *before* we began our lunar expeditions, and that the rush to get to the Moon was impelled by this mind-boggling knowledge? Was this the great impetus behind the expensive crash Moon programs?

Admittedly, this is only speculation, but we do know that our government space agency knew of studies that indicated that something was going on up there on our Moon—hundreds of unexplainable moving lights and objects, reported constructions, and unexplainable changes taking place on the surface of this strange, supposedly dead world.

Why did our government carry out and privately publish such a study? What did our space experts conclude about all this seeming UFO activity taking place on the Moon? Why did NASA itself commission Project Moon Blink, a program to search for lunar "events" which was carried out in cooperation with observatories around the world? Undoubtedly our space experts were looking for something. And at a time when our government and its military arms were busy trying to explain away or cover up the ever-increasing flood of flying saucers, it appears that possibly UFOs were using the Moon as a base of operations.

In addition, we now know that before we went to the Moon several studies, including a motion study of the Moon, did indicate that it could be hollow. Scientists in general agree that there is no such thing as a *naturally* hollow Moon; that if it is hollow then it was hollowed out artificially. Could not a few of our top scientists and government officials (also primarily military) not come to the conclusion that indeed it might be, which would mean that the Moon has to be a spaceship? And this knowledge further was a driving force behind the thrust to get to our Moon and check out this strange world once and for all.

Thus it would appear that, given these facts, there was much more behind the crash Moon program than our space agency or our government would have us ordinary citizens believe—certainly much more than a propaganda race with the Russians to get to the Moon first. Most Americans appeared to buy this story.

Dr. George E. Mueller, NASA's Association Administrator for Manned Space Flight summed up some of these "reasons why" that lay behind the space program in a statement to Congress which he made back at the outset of our lunar program:

"To the scientific world, there is great interest in the origin and history of the moon and its relation to Earth and to the solar system. Was it formed with the Earth, or captured later? Are there clues to the origin of life? To quote the President's Science Advisory Committee, 'Answers to these questions may profoundly affect our view of the evolution of the solar system and its place, as well as man's in the larger scheme of things.' "

Admittedly, this is true—the Moon may well, in fact,

affect man's entire outlook on his own origins and destiny! But were the feeble reasons he offered or even the supposed propaganda race with the Soviets the only reasons that lay behind our $25,000,000,000 effort to get to the Moon?

Not surprisingly another independent researcher also became intrigued by the Moon. He found revealing NASA photos and informed by an ex-NASA scientist willing to talk arrived at the same conclusion as we did. George Leonard in his mind-boggling book *Somebody Else Is On the Moon* states this thesis boldly: "The prime reason for the United States' launching an expensive Moon program was the recognition at official levels that the Moon is occupied by intelligent extraterrestrials who have a mission which does not include dialogue with us and may even be inimical to our long-range welfare."

Although we would agree with Mr. Leonard's overall conclusion, we are convinced from our own study of the moon that these lunar beings, who every evidence indicates have been around for thousands of years are in no way inimical to mankind. However we wholly agree with this conclusion of Leonard's that the U.S. and Soviet governments began Moon exploration because they were tipped off by their own scientific studies indicating that the Moon was more than just a natural satellite.

One problem arises even if we are to assume that all this is true. It may be understandable for the secretive Soviet government to keep all this under wraps, but with our open, aboveboard, democratic government, how is this possible? This is just like all the other secret maneuvers and escapades that Washington has pulled off, be it FBI or CIA that there is much that is secret within Washington's bureaucratic halls, as the past decade has proven. And there is every evidence also that secrecy swirls around our lunar program.

But again the question arises: Why?

Can you imagine what panic would ensue if the government were to acknowledge that UFOs stream into our skies every day? (Some, like Dr. J. Allen Hynek, astronomer at Northwestern University and former Air Force consultant on UFOs for twenty years maintain they number as many as six or seven every day!) Can you imagine the additional shock given to the public, if they were informed that aliens are using the Moon as a base of operations and have been for a long time?

And beyond that, can you imagine the shock of a stunning official announcement that the Moon is really a spaceship? What if the President of the United States were to get on television and announce to everyone that we now know beyond the shadow of a doubt that the Moon that circles our Earth is not a completely natural world but a huge hollowed-out planetoid steered into orbit around our world? Furthermore, that we now know that the Moon is inhabited, something our experts have suspected for decades?

Can you imagine the impact of such a shocking announcement on the world? The stock market would probably immediately crash. People might panic, perhaps even into a rash of suicides. An epidemic of disorder and chaos could sweep the country and the planet. The way people were agitated and alarmed by a radio program back in 1938 might be indicative of what would happen. Perhaps. But we doubt it. Despite all the U.S. governmental and scientific pronouncements to the contrary, official conclusions of university studies sponsored by the U.S. government show that the majority of people today—and the educated majority—are convinced that UFOs exist. Why, 15 million Americans are convinced they have seen a UFO!

So we doubt very much if people would panic. They, in our opinion, are ready for the truth—whatever it might be. As evidence that people can "live with it," we cite the fact that the human race lives every day under the shadow of the nuclear nemesis and there is no panic. And what about those millions who stay in California despite the threat from the best scientific sources that a devastating earthquake could rock and destroy their beautiful state tomorrow? They could panic and leave, but few have done so. So too we are convinced with the Moon. If the people were given the truth they perhaps would even welcome it, along with any alien "Moon beings" that might come down here and take over, as some science-fiction writers imagine. After all, they couldn't make much more of a mess of this Earth than our present leaders have already done, could they?

Seriously, we are convinced that we have on our hands a cover-up of cosmic proportions. A cosmic Watergate whose lid hides not just earthly indiscretions or corruption on the part of governmental officials and politicians, but a cover-up with Earth-shaking dimension and implications.

And this, as we have noted, is not just our conclusion but that of several leading investigators. Consider George Leonard's testimony. This professional writer who formerly held a number of federal posts in various governmental agencies as well as being an amateur astronomer, did a close study of the NASA photos—those that he could get his hands on—and came to the conclusion that the Moon gives clear evidence of artificial constructions. Further, he is convinced that the Moon is occupied by alien intelligence, and that NASA knows this but is carrying on what he calls Operation Cover-up.

Leonard was aided in his history-making search for the truth by a pseudonymous ex-NASA scientist, a man who one critic of his book *Somebody Else Is On the Moon* actually calls Leonard's "Deep Throat." This enterprising researcher and reporter had other contacts in the inner sanctum of our space agency, who have helped him expose the tip of this cosmic iceberg.

The conclusion of Leonard's government-shaking study: "The Moon is occupied by an intelligent race or races which probably moved in from outside the solar system. The Moon is firmly in the possession of these occupants. Evidence of Their presence is everywhere: on the surface, on the near side and the hidden side, in the craters, on the maria, and in the highlands. *They are changing its face. Suspicion or recogntion of that triggered the U.S. and Soviet Moon programs*—which may not really be so much a race as a desperate cooperation." (Emphasis added.)

But it might be objected that it is hard to conceive that the Russians and the United States have been cooperating in this secret venture in space. Isn't it interesting that, purportedly just to show goodwill, the U.S. and the U.S.S.R. carried out a 250-million dollar rendezvous in outer space (Apollo-Soyuz)? We know that such a link-up of American and Soviet space hardware required that they be made compatible with each other. Was such unusual cooperation in the paramilitary-space domain not an indication of a deeper hidden motive, possibly of future necessary cooperation in space that the knowledge of extraterrestrials out there watching us might impel?

Other hints of this space and lunar cooperation come readily to mind. Anyone who has researched the lunar space results becomes aware we have been exchanging information and data freely back and forth. That we would be so generous as to turn over our findings to them is understandable given the Santa Claus nature of our free-enterprise system and our generous U.S. government. But that the tight-fisted, secretive Communists are cooperating is interesting. Yet documents like NASA's Apollo 17 report contains references to such Soviet data and information which indicate that they have done just this—handed over vital information to us.

There is even more evidence. George Leonard notes that "a careful review of total U.S. and Russian space shots seems to indicate that a parceling out of responsibility has existed from the start." (*Somebody Else Is On the Moon.*)

Note this: *"From the start."* This goes way back to the late fifties, when we were still locked in the throes of the Cold War—long before the Nixon-Kissinger détente moves began. Intriguing.

Even more intriguing is the alleged meeting that George Leonard claims his ex-NASA contact told him took place in Europe in the spring of 1975. His source told Leonard that the advanced industrialized countries of the world met at that time. "The meeting was in England. They wanted to talk on the quiet about extraterrestrials and what they're up to. A lot of people at the top are scared."

Further, this ex-NASA scientist told Leonard: "They invited a physicist from Colorado, a man named Joachim Kuetner, who'd worked on the Moon program and knows what's up there. He could tell them about it first-hand. I don't know exactly what they talked about. But you can bet they know now that it's not Earth people's Moon anymore—if it ever was. It belongs to Them." (*Somebody Else Is On the Moon.*)

The implication here, of course, is that the Moon has been inhabited since the dawn of man. Leonard's source also told him that a number of top-level scientists inside and outside NASA are convinced that the Moon is actually a huge hollowed-out spaceship.

HOW MUCH DOES NASA KNOW ABOUT OUR SPACESHIP MOON?

George Leonard, whose research concentrated on activities and developments of alien beings on the outside of the Moon while we tried "to get beneath the surface of things," as he himself put it, makes an interesting observation concerning this vital question: "These data [that he himself garnered] have clearly shown up in a limited number of pictures made available by NASA, following a limited amount of study and research performed primarily by one person. What would a really systematic search of the Moon's surface produce? *How far has NASA gone?*" (Emphasis added.)

We echo Leonard's thoughts, for our work is the result of a limited amount of research, and if we alone, without ready access to the information uncovered by NASA and the Soviet space programs, uncovered this much startling information about our Spaceship Moon, what do NASA and the Soviet agencies, with all the data, all the evidence, and all the information garnered by their top scientists and engineers, know about our Moon?

Would not those revelations stagger the world if these findings were ever published? Would not we then have the final compelling proof that the world circling us is truly a spaceship—and an inhabited one at that!

Not for nothing did Santayana once contend that life is a movement from the forgotten into the unexpected.

—Loren Eiseley

SEVENTEEN

ANCIENT ACCOUNTS OF A WORLD INSIDE THE MOON!

Many gifted geniuses in the history of mankind seemed to have had the ability to see exciting facts of life and our universe that no person up to their time knew about. The great English poet Shelley, for instance, described the Moon in startlingly modern terms. He wrote that our satellite was once an orb "wandering companionless" through the cosmos "among the stars that have different birth."

This startling revelation comes in his poem "To the Moon," in the second stanza:

> Art thou pale for weariness
> Of climbing and gazing on the earth,
> Wandering companionless
> Among the stars that have different birth . . .

Exactly what the last international lunar conference concluded.

The question that should haunt every thinking mind: How could Shelley have possibly known this? Did he gain this insight intuitively or is it simply a good poetic guess?

It would seem that Shelley, who was steeped in the writings of ancient classics gained his knowledge from them. For Shelley is not alone in possessing startling insights into the truth of our satellite. Somehow many great men of ancient times guessed the truth about our Moon. There is even an entire host of ancient writers who insist that the Moon is a hollow world inhabited by living beings!

In fact, ancient writings are so filled with this startling reference that in 1644 an English bishop, John Wilkins, who was himself steeped in ancient classics wrote a book, appropriately entitled *Discovery of a New World,* in which he insisted there was a world of living beings inside our Moon! The good bishop (brother-in-law of the famous Oliver Cromwell of English history) did not arrive at this conclusion from the meager scientific evidence of his time but from innumerable references in the ancient works in which he was widely read. As a scholar of these works (mostly Greek and Latin), he had come across references time and again to the fact that our Moon was a hollow, inhabited world.

Finally, overwhelmingly convinced that these great ancient thinkers knew what they were talking about, the good bishop decided to collect together this evidence in a book. He came to the conclusion that the Moon was indeed not a solid orb. For as he says in his introduction: "If there be a habitable world in the moon (which I now affirm) it must follow that her orb is not solid as Aristotle supposes."

Who does Wilkins cite as possessing this esoteric knowledge? For one Xenophanes, renowned student of Socrates, who, as he put it, "conceived the moon to be a great hollow body, in the midst of the whole concavity, there should be another globe of sea and land, inhabited by men as our earth."

Fantastic? Hold onto your exclamations! This is just the beginning.

THE GREAT REVELATION OF ORPHEUS

Perhaps the most astounding quote that Bishop Wilkins makes comes from the great Orpheus, one of the most ancient of Greek poets, who, interestingly, some claim was the son of the Greek god Apollo, after whom the entire manned Moon landing and exploration program was named!

This god (whom many modern historical scholars, as we shall see later in this chapter, now consider might have been a real person, and who Aeschylus, the father of Greek drama, tells us could "calm the seas") purportedly was the author of the ancient Orphic hymns, in which he is thought to have transmitted much knowledge, much of it accurate pieces of information, especially in the field of astronomy, that turn out to be astounding. His other revelations are equally so.

Speaking about the Moon, Orpheus, the Greek god of wisdom and father of the wondrous Greek culture, tells mankind "that it hath many mountains, and *cities and houses in it*. . . . To him assented Anaxorgas, Democritus and Heraclitus . . . All who thought it to have firm solid Ground, like our Earth, *containing in it many large fields, champion grounds and divine inhabitants!* This belief of Orpheus was related by Plutarch, Diogenes and Laertius." So relates Bishop Wilkins. (Emphasis added.)

In light of the Spaceship Moon theory, utterly astounding!

Before we get to some other startling insights revealed by Orpheus "the Moonman," as he was called, let us consider his background. Appropriately, as we have seen, the entire Moon landing program and its individual space expeditions have been named after his father, one of the greatest gods of antiquity. A few scholars claim that Apollo, who was the twin brother of Artemis, goddess of the Moon, was in actual fact an ancient astronaut.

Could it be possible that this god of prophecy, who held sway at the temple of Delphi, this god of poetry, healing, and light, and his enigmatic son and all the other gods who lived on Mount Olympus were actually alien beings from beyond this planet? Of course, there is no way of proving this intriguing speculation, and without proof it is ridiculous to consider it, as some sensationalistic writers do, "absolute fact."

But we do know that the Orphic hymns, which ancient tradition tells us this Greek god of wisdom authored, do contain much startlingly advanced astronomical knowledge. Advanced, that is, for such an ancient work.

A contemporary astronomer, C. S. Chassapis, analyzed the Orphic hymns and came to the conclusion that the ancient Greeks of the second millennium (2000 B.C., or 4000 years ago!) indeed did have an advanced knowledge of astronomy. Chassapis claims that the author of these

Orphic poems, whoever he was, knew that the seasons were caused by Earth's rotation around the Sun along the ecliptic, and knew that the torrid, temperate, and frigid zones existed on our planet.

Orpheus is generally considered to be the author of these works. Chassapis points out that the work itself indicates that Orpheus could figure the solstices and calculate the equinoxes. He must have also known that Earth's rotation on its axis causes the apparent rotation of the Sun and the stars.

Furthermore, Chassapis also points out that the second-milennium Greeks used a calendar of twelve months, which they calculated from full moon to full moon. And they seemed to know that all phenomena are governed by universal cosmic laws. Tradition, which tells us that Orpheus revealed all this in his hymns, holds that Orpheus is responsible for revealing the fact that the world is "egg-shaped," and he is given credit for being the ancient Greeks' source of knowledge about mountains on the Moon!

All this is startlingly advanced knowledge for mere backward Earthlings who supposedly just crawled out of caves! But there is more that this "god-man" revealed to humans which really rocks us back on our heels and makes us wonder if indeed Orpheus, along with his father, Apollo, and the other gods, was not a myth from the planet Earth but a real being alien to this world—an ancient astronaut. Orpheus also revealed to mankind that man's nature is dual: that he is in part of the Earth, and in part of the heavens.

In one revealing, mind-blowing passage this Moonman, this son of Apollo, tells us something that before the facts were known about our Moon meant nothing—but in light of the Spaceship Moon theory means everything. Orpheus supposedly uttered this revealing statement about his fellow heavenly gods: "These innumerable souls they fell [traveled] from planet to planet, and in the abyss of space, lament the heaven they have forgotten."

These startling, stunning words are astounding, for they seem to imply the existence of intelligent life on other worlds and that advanced space beings came to our planet from another corner of the universe. They are particularly revealing in light of Spaceship Moon, which he tells us is hollow and "has many cities and divine inhabitants" inside it.

Since so many Greek gods were connected in one way or another with the Moon, could not Orpheus be referring to the very beings who came into orbit around the planet Earth in this Spaceship Moon? And was not Orpheus implying that these "gods" who came from this Moon had actually traveled great distances through space for eons? That they longed for that corner of the cosmos from whence they came?

Mind-boggling! We do know that one Greek myth informs us that Orpheus the Moonman was supposedly slain by Zeus, head of the gods, "for divulging divine secrets." After his mortal life was over, the myth continues, Orpheus returned to the Moon.

With Orpheus' words haunting our minds, we come to ponder anew that question of questions: Is this mere myth or is there reality hidden here? Was Orpheus a real being and was he not indeed revealing facts about our mysterious Moon?

Very few scholars today, of course, would hold that these Greek "sky-gods" were in reality spacemen—superior beings from another planet. Such science writers, however, as the British editor Brinsley La Poer Trench, and French author Jean Sendy, as well as the late Otto Binder, former NASA science researcher, do claim that they were ancient astronauts. Obviously, no ancient Greek held this wild theory or was even aware of such a possibility. However, unquestionably the ancient Greeks, even the greatest thinkers like Aristotle, Plato, and Pythagoras, and yes, even Socrates, never really doubted that the "gods" existed—that they were real beings. In fact, Pythagoras and Socrates believed that they also received special inspiration from these beings.

Did the alien culture bearer who spoke about the stars and taught mankind a great deal of advanced knowledge beyond the ken of Earthlings of that time actually exist? The historian Will Durant holds that he "very probably existed, though all that we know now of him bears the marks of myth." (*The Life of Greece,* Simon & Schuster, 1966.)

The remaining quotes of the ancients that Bishop Wilkins gives in his startling book *Discovery of a New World* seem tame by comparison. But nonetheless shocking. For Wilkins tells us that Pythagoras, the great ancient Einstein/da Vinci of the ancient Greek world, "did affirm that the Moon is Terrestrial and that she is inhabited as this lower world [Earth], that those living Creatures and Plants which are in her, exceeding any of like kind, with us in the same Proportion, as their Days are larger than ours by 15 times." (P. 79.)

This too is a remarkable revelation, for how could the ancients have known that the Moon had a two-week-long day? It is actually just about about 15 times longer than ours. How could Pythagoras have known this? We do not know, unless he was privy to other revelations of the "gods" themselves. It is known that Pythagoras, considered the founder of Greek mathematics, had spent twenty-two years in Egypt, where he was in the inner sanctum of highly knowledgable priests who were supposed to be privy to the esoteric knowledge of the ancients and even of the "gods."

Similarly, Wilkins tells us Plato agreed with Pythagoras. As the good Bishop informs us: "To this opinion of Pythagoras Plato also assented, when he considered that we may often read in him, and his followers of another Aethera Terra, and lunes populi, that is an Athereal Earth, and Inhabitants of the Moon."

Plato, like Pythagoras, spent considerable time traveling throughout Egypt. After the death of Socrates he joined the secret Pythagorean societies of the time and was introduced into the inner sanctum of esoteric knowledge.

Maybe this accounts for Plato's strange philosophy of a world of ideas and his belief that thinking is essentially remembering and that the real world of ideas exists not on this Earth, which is merely a "shadow world," but elsewhere. According to Wilkins, that other "real world" was the world of the "gods" inside the Moon!

Thinking is essentially remembering? Perhaps it refers to remembering back to a time when, as all mythologies relate, "the gods" came down from the heavens and freely communed with man. Maybe, as Leonardo da Vinci points out, Plato was right in asserting that man knows all things but has forgotten most.

These are admittedly strange, startling thoughts. But no stranger than the idea behind them or the thoughts and knowledge that generated them, the mysteries of Moon as related by the ancients. For Bishop Wilkins, the classical scholar, astounds us with these revelations.

Discovery of a New World, which reveals what the ancient "gods" revealed about the Moon—that it is a hollow world with cities inside of it and is indeed the home of "gods"—however, does not deal with ancient revelations only. Wilkins claims that some medieval thinkers in his own time also held that the Moon was an inhabited world. Even great scientists as Kepler, along with Copernicus, Wilkins insists, held that Moon was filled with living beings. In Kepler's case, according to Wilkins, he held it was inhabited on the inside!

Wilkins says: "Kepler calls this world by the Name of Levania from the Hebrew word meaning Moon and our own Earth-revolving world by reason of its Diurnal revolution appears unto them constantly to turn Round, and therefore he stiles (*sic*) them who live in that Hemisphere which is towards us, by the title of Subvolens, meaning they who revolve under because they enjoy the sight of this earth."

It might be objected that Kepler only put forth these ideas in his science-fiction novel *Somnium* (*The Dream*). Not according to Wilkins, for he asserts: "And Kepler did not jest . . . indeed he protests that he did not publish them either out of Humor or Contradiction . . . or from a desire for Vain-glory, or in a Jesting Way, to make himself or others merry, but after a considerate and solemn reason, for the discovery of the Truth!"

Wilkins also asserts that "Copernicus affirms this Hypothesis along with the great ancients Aristarchus and Philolus." Not only does Wilkins hold that Kepler claimed the Moon to be inhabited, but at least one writer—former NASA researcher Otto Binder—speculates that the space journey to the Moon Kepler described might have actually taken place. Or at the very least spacemen who made the trip imparted this knowledge to Kepler.

Why would Binder make such a preposterous claim? Simply because the knowledge contained in that science-fiction fantasy is "astonishingly accurate in details." Kepler wrote of principles of space flight that we know today to be established fact—the shock of acceleration, weightlessness of the body, free fall in orbit, in addition to amazingly accurate descriptions of spacesuits that had to be worn by crews of the spacecraft. Dr. Clifford Wilson, noted scholar, investigated Binder's claims that he set forth in his book *Unsolved Mysteries of the Past* and claims "it is possible that Kepler's terminology and knowledge could to a great extent be attributed to his own capacity and foresight as an astronomer." How else explain such knowledge back in the 1600s?

Although most literary and scientific scholars would most certainly dispute Wilkins and Binders's interpretation of Kepler's book, which was really the first lunar guidebook, simply because it was written in a pseudo-scientific, partially mystical manner, he did people his Moon with strange beings who lived underground in huge caverns, the kind of caverns that we have pointed out some scientists speculate might exist inside the Moon.

Samuel Butler, the English poet, wrote of Kepler's conception:

> ... Th' Inhabitants of the Moon,
> Who when the Sun shines not at noon,
> Do live in Cellars underground,
> Of eight miles deep and eighty around.

What about John Wilkins, the English bishop himself? What does he believe about the Moon? Unquestionably his was a serious book which held the Moon to be inhabited inside. He obviously also agrees with the great ancients and claims that "there are high mountains, deep Vallies and spacious plains in the body of the Moon."

He also predicted that someday man will ferret out the truth. "That tis probable for some posterity, to find a conveyance to this world, and the Inhabitants there to have commerce with them. Tis the Opinion of Kepler that as soon as the art of flying is Found out, some of their

Nation [Germany] will make one of the first colonies, that shall Transplant into that other World."

Wilkins's final proposition states: "That tis probable that there may be inhabitants in this other world but what kind they are is uncertaine (*sic*)."

This statement also bears up under the latest facts.

Was Wilkins "loony"—a lunatic or crackpot? Hardly. For, as Walter Sullivan points out in his book *We Are Not Alone,* "Wilkins was one of those men of that period who bristled with ideas." He helped form a society of savants which explored through discussion and speculation many of the new exciting avenues of direction that science and knowledge was taking at the time. These weekly meetings, often held in Wilkins's own quarters, included such greats as Robert Boyle, famed for his law on the compression of gases, Sir Christopher Wren, outstanding architect of St. Paul's Cathedral in London, and Samuel Pepys, well-known English diarist. Even Newton joined this distinguished group later. And that group's name? It became in time the famed Royal Society, which Walter Sullivan calls "one of the most distinguished associations of scientists ever formed."

> *Nothing but the admission of life and intelligence inhabiting the space around the earth-moon, binary-planet system will provide the UFO data with a common denominator of explanation and rationality.*
> —Morris Jessup

EIGHTEEN

THE MYSTERIOUS ALIEN SATELLITE
OF OUR MOON!

When man went to the Moon, many scientists secretly hoped and some fully expected to find evidence that extraterrestrial beings had been there before us. One such scientist who is convinced that such alien structures might be found is the British space expert V. F. Foster: "In reality such artifact devices may well embody the techniques and principles of superhuman knowledge. Almost certainly, we will soon encounter these objects on the Moon!"

The hopes of man sometimes are translated into fantasy writings. And as some philosopher has said "the wish is the father of the thought". Scientists had longed to find evidence to prove that man is not alone in the universe. It was not surprising that the great science fiction writer Arthur Clarke came up with a very popular book and film *2001: A Space Odyssey* which projected the discovery of such an alien artifact—a huge monolith buried on the Moon that emitted strong radio waves when discovered.

Now amazingly it appears that the Soviet Union's space probe has discovered a startlingly similar artifact. For a UPI news report based on a Soviet government announcement claims that on February 14, 1973 a Russian remote controlled space robot (Lunokhod 2) while probing the Taurus Mountain region (the same region ironically where we sent our last Apollo probe Apollo 17) discovered an unusual monolith, a smooth stone slab about a meter long, which remarkably resembled the smoothly carved stone monolith much like the one in *2001: A Space Odyssey*. The Soviet government report (announced by Tass) indicated it was like a sculptured piece of stone, "a plate . . . (with) a smooth surface".

The full UPI news report stated:

- The Lunokhod 2 moon robot parked just over a mile from the Taurus Mountains Wednesday (February 14) and probed an unusual slab of smooth rock blasted into view by a large meteor, the Tass News Agency said.

- This one-meter long plate has proved to be a strong monolith. The eight-wheeled robot, which arrived on the moon January 16, was nearly three miles from its landing site on the Sea of Serenity.

- The plate has a smooth surface, whereas giant stones lying nearby are pockmarked with holes of craters left by tiny meteorites, Tass said.

If such a discovery has actually been made, it is mind-boggling. First, Jules Verne in the nineteenth century described man's first landing on the Moon and was uncannily correct, not only on the point of departure (the coast of

Florida about 50 miles from the actual Apollo blast off); not only on the exact number of crewmen (three) but even with the correct speed the rocket-like projectile had to reach to break the bonds of earth's gravity. Now it appears that science fiction again has mirrored a future happening. It almost makes one wonder whether or not someone is out there watching us and leading us on. . . .

But, as shocking as this discovery is, another is even more profound. It too, eerily, was predicted by Arthur Clarke. Actually, Clarke's *2001: A Space Odyssey* novel was based on an earlier short story ("The Sentinel") which told how a similar device was purposely planted by an intelligent race of aliens to enable them to learn about the progress of man. And when discovered and examined by man, it touched off a series of prearranged messages via radio signals which were automatically sent to another part of the universe.

This kind of automatic warning system would enable them to monitor not only man's presence on the Moon, but in turn would let man know that an alien intelligence had visited this part of the universe. According to some scientists who speculate along these lines, such a device could be planted that could conceivably allow these distant beings to monitor Earth broadcasts, both radio and television, and thus enable them to keep a close watch over our progress (or lack of it).

THE MOON HAS SUCH A SENTINEL!

Believe it or not, a similar device of this kind has been discovered, a few scientists claim. Only it isn't a monolith buried in the Moon's surface but an unmanned "probe" robot satellite placed in orbit around the Moon thousands of years ago.

A Scottish astronomer named Duncan Lunan claims that he discovered such a robot satellite that was placed in orbit around our Moon. And he insists he has even translated its message to mankind. These startling claims were recently published in a research paper by Lunan in *Spaceflight*, the journal of the prestigious British Interplanetary Society.

This society is an international scientific organization of astronomers and other scientists which is headquartered in London.

Leonard Carter, the executive secretary of the British Interplanetary Society, claims that Professor Lunan discovered this unusual satellite after studying "radio echoes that have been known since the 1920's, but couldn't be explained as having originated from earth." (D. A. Lunan, "Space Probe from Epsilon Bootis," *Spaceflight*, April 1973, 15:4, pp. 122–33.)

In December 1927 a young Norwegian specialist on electromagnetic waves, Professor Carl Stormer, learned that some American researchers had received strangely delayed radio signals seemingly emanating from around the Moon. Stormer teamed with a Dutchman, Van der Pol from the Philips Research Institute in Einhoven, and in September 1928 began a series of experiments. They radiated radio call-signs of different lengths at 30-second intervals.

Within three weeks they received these signals back, but with delays from 3 to 15 seconds. The radio signals were registered with these delay intervals (in seconds): 8–11–15–3–13–8–8–12–15–13–8–8. This radio phenomenon was received again all through February 1929. These strange radio echoes were reported by researchers all over the world.

On October 24, thirteen days later, and on October 28, 48 more signals were received.

In August 1929 Professor Stormer published his findings. Almost immediately theories began to spring up trying to explain how the delayed shortwave impulses occurred. Could they perhaps be somehow reflected radio waves, possibly somehow reflected back from the stars? This ridiculous explanation was rejected along with the obvious possibility that they were reflections from the Moon.

While scientists continued to ponder this problem, the radio signals continued to be received through the 1930s and into the forties.

In the early sixties Professor Robert Bracewell of the Radio Astronomy Institute of Stanford University theorized that the signals could be a message from alien intelligence in outer space. Bracewell observed that if alien beings wanted to get in touch with us, they might use the delayed return of signals. The radio signals with delays perhaps did contain a message! Perhaps someone or something out there, according to Bracewell's suggestion, was trying to send us a message by trying to draw, as it were, a space picture with this series of delayed radio signals. Professor Lunan of Scotland, who was in communication with Bracewell, took the suggestion seriously and set to work trying to figure out this conundrum. He got hold of some of the original echo data, although unfortunately much of it appears to have been lost.

In 1973 the British Interplanetary Society's publication *Spaceflight* published Lunan's claim that an artificial satellite was beaming a message via delayed radio signals to Earth. According to the executive secretary of the learned and respected society, Leonard Carter, they did this to give Lunan's unbelievable claims an airing and to stimulate scientific work in this area.

Lunan announced in the British Interplanetary Society's journal that he was convinced that the radio signals were emitted from an artificial satellite that was placed in orbit around the Moon by unknown alien beings about 12,600 years ago. He claimed a computer on this satellite was preprogrammed so that it responded to radio waves from our planet whenever its own position in relation to our Earth was suitable for reception. Our radio signals were recorded by the satellite and sent back on the same wave length, only with "intelligent delays" to communicate a message which eventually intelligent man on the planet Earth would recognize!

According to Carter, spokesman for the British Interplanetary Society, "Lunan plotted the echoes on a graph. Oddly they seemed to make a series of dots outlining the [known] constellations. But they were slightly distorted. . . ." However, Lunan has gone into the question of this distortion and alteration. And the dots related to the constellations as they were about 13,000 years ago.

Carter then relates that Lunan became convinced from his studies of this particular constellation that this satellite robot was put into orbit by inhabitants of an alien planet which orbits a sun of the star called Epsilon Bootis.

However, Carter notes that Lunan, even after discovering this time lag in the echoes, found that the star Epsilon Bootis was still out of position on his graphs. Lunan also noted that a series of dots resembled no constellation at all, so he concluded that these really represented echoes of various durations, and really contained a message.

Lunan's research report indicates that computers on this satellite robot probe transmit the message whenever they are triggered by radio waves sent from Earth at a certain undetermined frequency.

This robot satellite placed in orbit thousands of years ago lay dormant, circling the Moon, until the 1920s, when men on the planet Earth began sending radio waves into space. This triggered the device, which has been sending a message to mankind ever since. The Scottish astronomer now believes he has translated the message. Here it is:

> START HERE. OUR HOME IS EPSILON BOOTIS, WHICH IS A DOUBLE STAR. WE LIVE ON THE SIXTH PLANET OF SEVEN. CHECK THAT—THE SIXTH OF SEVEN COUNTING OUTWARD FROM THE SUN, WHICH IS THE LARGER OF THE TWO.
>
> OUR SIXTH PLANET HAS ONE MOON. OUR FOURTH PLANET HAS THREE. OUR FIRST AND THIRD PLANET EACH HAVE ONE.
>
> OUR PROBE IS IN THE POSITION OF ARCTURUS, KNOWN IN OUR MAPS.
>
> OUR PROBE IS IN THE ORBIT OF YOUR MOON.

Lunan's work certainly is intriguing. But does his interpretation have validity? Remember, as we have noted, this interpretation is based on the unexplained radio echoes heard in the late twenties and early thirties, detected by French, Dutch, and Norwegian radio researchers who at the time were transmitting a series of telegraphic code broadcasts. Only they received back *two* sets of echoes. They noticed that it took one seventh of a second for the return of the first echo, the exact time it would take to bounce a radio wave off the ionosphere.

Lunan calculated that "because a second set of echoes came back after delays of three to fifteen seconds, the messages could have been intercepted, interpreted and then rebroadcast by an object of intelligence circling the Moon."

Once it dawned on Professor Lunan that the delays could be coded messages, he set to work and translated the message.

In a short time Professor Lunan made his discovery. "To my astonishment," Lunan declares, "the dots made up a map of an easily-recognized constellation—the Constellation Bootes in the northern sky. The curious pattern of delayed echoes was a pattern of star positions."

Lunan believes the alien beings who set this satellite into orbit around the Moon were from the star Epsilon in this constellation. This star, astronomers now know, is one that has exhausted the reserves of hydrogen in its core and is contracting and growing hotter. Any intelligent life on any of these planets circling this dying star would have to leave this solar system altogether in order to survive. If we were in such a position and had to leave our sun, this would be the best way to do it: Dandridge Cole and Isaac Asimov's conception of hollowing out a nearby asteroid or planetoid, filling it with the necessary supplies and accouterments of their civilization, and setting sail on a powered, self-contained, self-recycling enclosed world away from their dying star and into the endless oceans of space of the outer universe.

Perhaps this is what drove the beings who built the lunar spaceship world now in orbit around the Earth. Is there any connection between the two? It is intriguing that Lunan calculated that the alien beings who placed their robot satellite into orbit around our Moon did so 12,600 years ago. This figure is strikingly close to the figure given by Bellamy, who calculated that the Moon came into orbit around our Earth between 11,500 and 13,500 years ago! (See *Our Mysterious Spaceship Moon*.)

Could it be that these alien beings who steered the Moon into orbit around our planet were the same beings who left this artificial satellite in orbit around our Moon?

Although the findings of Duncan Lunan, president of the Scottish Association for Technology and Research in Astronautics, have yet to be verified, their implication is staggering. One scientist commenting on Lunan's findings (who claims that the chance of different radio echo delays forming star maps purely by coincidence would be 10,000 to 1) maintains this satellite might be a highly sophisticated computer with an enormous store of information ready to be imparted to mankind! He strongly believes that if this alien space probe is confirmed, concentrated effort should be made to "interrogate" it.

It is precisely for this reason that a group of British scientists have begun a serious attempt that could prove to be one of the most fateful and profound in the entire history of mankind.

How can this be done? R. N. Bracewell of the Radio Astronomy Institute, Stanford University, who originally suggested that an unmanned space probe might be used by aliens to beam a message via delayed radio waves, according to Lunan claimed: "If we returned the signals to it again, it would know it had established contact with intelligence...."

"Should we be surprised," Bracewell writes, "if the beginnings of its messages were a TV image of a constellation?"

Lunan suggested that not only radio probes be used but says "attempts might be made to contact the spacecraft by laser probing...." (*Spaceflight,* April 1973.)

He also suggests radar attempts.

Although Lunan's theory is far from proven, scientists are planning to put it to the test. Anthony Lawton, a British computer specialist, said an experiment will send radio signals to try to stir the supposed satellite into another response.

What is the opinion of other scientists connected with this tremendous discovery? Do they believe that Lunan's interpretation is correct?

Professor Ronald N. Bracewell of Stanford University, one of America's leading radio astronomers, admits he has been exchanging research material on this matter with Lunan for the last few years. He does not discount Lunan's interpretation of the radio signals, but has admitted that he has reservations about them.

As Bracewell explains: "If he's right in his thinking this material did contain a message. However, Lunan's representation of the dots as the constellation of 13,000 years ago may not be quite right. It could be simply because this is not a message at all."

Leonard Carter of the British Interplanetary Society agrees with Bracewell that in fact Lunan's interpretation might not be correct. He said that his society published Lunan's research findings "so that any scientist with records of the mysterious echoes can bring them to the attention of scientists working in the field."

Concludes Carter: "There will have to be a complete reading of all this material to check all the echoes."

"They [the echoes] exist in fact. When one plots them as Lunan did, one gets a very, very curious result. It's in the interpretation of them that one let's one's hair down. It could mean many things."

Lunan is convinced his work has been verified. Additional researches he has worked out have produced more star charts based on the delayed echoes. Lunan now has come up with six different star maps. By studying them all together he interprets that "every one of the reference lines points to a star called Epsilon Bootes," which he believes is the point of origin of this alien Moon satellite.

However, another researcher, Anthony Lawton, in his book *Ceti* (Warner Books, 1976), claims that Lunan is wrong. First of all, he believes that he has solved the puzzle of "the Long Delayed Echoes of the Twenties," insisting that they were "not received in the sequence Lunan used them to compile his star patterns—and, in any case, many other star systems could have been made to fit the same sequence." So states Lawton.*

Furthermore, Lawton claims that "they were not from an alien probe" but that "they are a perfectly natural phenomenon."

Whether or not Lunan's interpretation is correct, two

---

* Lawton believes that the LDEs (Long Delayed Radio Echo Signals) are "caused by signal reflection from the upper atmosphere."

He does admit: Some researchers might now suggest that an alien probe could well be using the Lagrange area (around the Moon) as an Earth-orbiting "parking lot" from where it is sending us signals. He objects, however: "This is unlikely because the power needed by such a probe would make it a very conspicuous object." This appears weak, for even given the rapid pace of technology that can put power packs in minute packages, what could aliens millions if not billions of years ahead of us accomplish?

things are certain. First, that radio signals have been detected in the vicinity of the Moon by various radio researchers at various times. In 1927, 1928, and 1934 a series of mysterious signals was intercepted on the Earth. More radio signals came through in 1935, a series of mysterious signals intercepted on Earth by other scientists. And in 1935 scientists Van der Pol and Stormer did pick up echoing radio signals around the Moon. They calculated that according to the time lag the signals seemed to be reflected from a large object about twice the distance that Earth is from the Moon, or about 500,000 miles away!

As this radio phenomenon continued to occur, other scientists in the past decades narrowed the object's position to "an area 60 degrees behind the Moon as the lunar world orbits the Earth. This object could be the robot satellite that Duncan Lunan believes is emitting the strange alien message." It is possible that it could even be a different one very much like it.

Science writer Joseph Goodavage claims: "The object trailing the Moon at twice the Earth-Moon distance has so many peculiarities that some scientists are sure it's artificial. It would be impossible for a large asteroid to maintain a relative [geocentric] velocity equal to the Moon at that distance." (*Saga,* April 1974.)

If this object is actually the Lunan satellite and his message is accurate, we are still left with this tremendous fact to ponder: Epsilon Bootes, the star from which our alien visitors purportedly came, is 103 million light-years away —well beyond our Milky Way Galaxy!

For a cosmic spaceship to make that tremendous journey, even at the speed of light, would require 103 million years!

As Goodavage concludes: "Nothing we know of, and certainly no living creature or organization of entities that we can imagine, is capable of cohesion for such a vast period of time." (*Saga,* April 1974.)

But then again, the alien beings of this Spaceship Moon might be. Man, if we are to believe scientists of Earth, has been on our own spaceship planet for the past few million years, making tediously slow progress. Now man has taken his first feeble infant steps into space. And he has reached the Moon.

Already some men of Earth have conceived the scheme

of hollowing out an asteroid or planetoid and converting it into a completely self-contained, self-recycling, self-sustaining world to journey through the cosmos to the stars. A journey that theoretically could last millions upon millions if not billions of years. And even if and when man's first self-made spaceship world should wear out or be damaged beyond repair, he could simply lock himself into orbit around another large asteroid or planet with a good-sized moon (as our own Spaceship Moon tenants might have done) and fashion it into a new world.

Perhaps this is what will happen to man someday. Perhaps thousands or even mere hundreds of years from now man may find himself hurtling through the universe in a similar spaceship, journeying through the endless reaches of space. And, who knows, someday, somewhere. some of man's ancestors will be puttering around some world on a satellite similar to that which alien beings may have left in orbit around Earth, complete with a message to some unknown beings. And we might be leaving a map of our star system for some intelligent beings to figure out.

> *Have we not lately in the Moon*
> *Found a new world to the old unknown?*
> *Discovered seas and lands Columbus*
> *And Magellan could never compass?*
> —Samuel Butler

NINETEEN
THE MOON: THE CHALLENGE OF OUR TIMES

Shortly after the last Mercury space mission was run, NASA received an interesting letter from a motor mechanic in Milwaukee. It read: "I see from my newspaper today that the Mercury program cost 2 dollars and 6 cents for every man, woman and child in the United States of America. I enclose a mail order for $2.06, and I'd like to see the whole thing over again."

We would like to see the Apollo Moon missions run again. We would like to go back once more to take another close look, this time with our eyes wide open and with an open press policy for our space agency. However, this is unlikely to happen for a long, long time—if ever. Man may go back someday but it is doubtful that he will for another decade or more.

Originally ten flights to the Moon had been scheduled, two more after the epoch-ending Apollo 17. In fact, Apollo 18 and 19 rockets were already paid for and the astronauts trained and raring to go. Yet these last two scheduled Moon missions were cut. Why were they slashed when, as one science reporter tells us, "the hardware had been already purchased"?

Supposedly many people in our country were screaming against this expensive boondoggle, which they critically called a "Moondoggle." So Congress purportedly slashed the program. But the truth is that after tens of billions had been spent the cutback itself saved only $20 million per flight in operational costs since the rockets and space hardware to be used had been already paid for and were being readied to go. Even critics of manned space flights like Dr. Thomas Gold of Cornell University screamed out against those nonsensical cuts: "It's like buying a Rolls-Royce," carped Gold, "and then not driving it because you want to save a few bucks on the gas."

The *New York Times* editorial of September 4, 1970, lamented this regrettable decision to slash Apollo Moon flights 18 and 19, observing that "an incredibly intricate technology and elaborate organization built to exploit that technology are, in effect, being abandoned . . . now that the easily bored world audience has begun to yawn."

Although ostensibly the move was motivated by budget concerns, is it not possible that our space agency already had conclusive evidence that the Moon was not only a spaceship but a presently occupied spacecraft? Did the astronaut encounters with UFOs worry NASA officials that another trip might bring startling encounters or revelations that might blow their cover of secrecy?

Ostensibly NASA space officials and certainly many leading lunar scientists around the world would be eager to return. Earth scientists have pointed out time and again that trying to unravel the make-up of the Moon by sampling a half dozen areas of her surface is an extremely difficult job. As one lunar expert testified: "I would suggest that discovering everything we would like to know and need to know about the Moon in six landings is somewhat tantamount to trying to describe what the North American continent is by sampling it in six spots for a few hours each."

Six manned trips to Luna have, if we are to believe NASA, settled nothing. Their scientists freely confess: The major questions remain. In fact, more questions, more contradictions, more confusion now exist than before Apollo started, some of them say. All we can be sure of is that the Moon swirls with mysteries. This may be true, but, as we have seen, the scientific facts indicate that she is, as our two Soviet scientists theorize, an articially hollowed-out planetoid steered into orbit around our water world.

The characteristics of the Moon's make-up indicated that even before we went up there. NASA's Apollo data and findings tend to prove this. Even ancient documents and histories as well as legends and myths tell us this is the unbelievable truth.

We also realize, however, that it could be wrong. But if it is, it would leave us with that bundle of contradictions, conflicts, and inexplicable mysteries, unexplainably accurate historical accounts, and, more importantly, inexplicable scientific facts that would then seem to be totally incomprehensible.

Many space critics before the Apollo program complained "Why go to the Moon?" History, however, has proved that unexpected rewards often come from exploratory endeavors. For instance, Columbus's voyage led to the discovery of a new world. Now it looks like our Apollo Moon voyages have led to the discovery of a new world that man never dreamed existed.

Columbus's great exploration bore fruit even though he died, criticized and unsung in prison chains, without the great realization that he had in fact discovered a new world.

In truth, it has been said of Columbus that (1) he did not know where he was going (which was right, since he thought he was going to India); (2) he did not know where he was when he got here (true again, for he thought he had reached India and even called the natives here Indians); (3) Columbus died never realizing where he had been!

So too with our Apollo program. Our astronauts and astronomers never realized, in a sense, where they were going. Everyone—except maybe a few in the inner sanctum of our space agency—thought we were journeying to Earth's natural satellite. Actually, our astronauts went to an artificially created alien world—a spaceship in our skies that was steered into orbit around our planet. They may have even gone to a presently inhabited world, if we are to believe the UFO sightings they experienced while up there.

Now it appears today that they really did not know actually where they had been—at least if you are to believe NASA. Will the astronauts die never realizing what kind of a world they had really traveled to? Personally, I am convinced they already know.

MAN'S IMPOSSIBLE DREAM

The Moon at one time seemed unreachable—an impossible dream. But man has achieved the unattainable, has reached the unreachable, and in so doing has given a tremendous lift to the sagging human spirit in these chaotic times. It should be an inspiration to thinking men everywhere to reflect what the foresight and energy of a people determined and dedicated to a goal can accomplish. The great space program culminated at the end of the sixties, as President John F. Kennedy foretold, with man's exploration of the Moon—the single greatest achievement of modern times and without a doubt the single greatest accomplishment of all time!

Now it turns out that our lunar space program may very well also be the single greatest discovery of all time—for it has revealed the truth about our neighboring world, a revelation that may well lead to a better understanding not only of our own world but even of ourselves—even of our origin and our destiny!

Should man therefore go back and try to remove all doubt? This, of course, is debatable. In the sixties before man went to the Moon, Bertrand Russell wrote an article which was published in the *London Times* on the eve of the Apollo 11 Moon launch. The article was entitled "Why Man Should Keep Away from the Moon." For reasons Russell never dreamed, perhaps he is right. Perhaps man's going to the Moon is a flirtation with undreamed-of disaster.

But we doubt it. More probably it will lead to definite proof of undreamed-of discovery.

A reader of my first Moon book pointed out to me that a well-known American psychic, David Bubar, Baptist minister and executive director of the Spiritual Outreach Society (SOS), an organization devoted to psychic research, predicted at the outset of our Apollo voyages that "within the next 60 months (by 1975), we will discover that a highly developed form of life exists in outer space and has existed for a long time. . . . Coupled to this will be further study of the origin of the Moon that will reveal facts so shocking that we will think twice before venturing any further into the deep blackness of space!" (René Nooriberger, *You Are Psychic: The Incredbile Story of David Bubar.*)

Would reopening our lunar space program open a Pandora's box of problems? Possibly. But we hold this to be a false presumption. First of all, if alien beings really do exist in the Moon, they certainly do not mean us any harm. Evidence shows they have been around a long time, as we have seen.

NOT THE REAL MOON AFTER ALL?

The great Apollo voyages to the Moon are over. These Moon trips were a mind-stretching leap into the cosmos. It turned out to be a journey into the unknown—a journey to a world of mystery that man never really understood or expected ever to exist. Richard Lewis, that science reporter par excellence, summed it all up so well: ". . . perhaps the Moon the scientists thought they were seeing was not the real Moon at all, perhaps they were being fooled, as the early sixteenth-century mariners were when they believed that Newfoundland was a promontory of the coast of China." (*The Voyages of Apollo.*)

The strange world of our Moon, which we have so recently explored, appears to be a strange new world for man—the strange world of our Spaceship Moon.

YOUR MOON ODYSSEY

Undoubtedly you the reader set out on this Moon odyssey as a pure skeptic—as I myself did. After all, man has been studying our satellite for centuries—if not for thousands of years—and no scientist up until recently has ever suggested that the Moon is anything more than a huge natural satellite circling our Earth. Even though some more enterprising scientists wondered about those reports of strange moving lights and happenings taking place on her surface, and a few even speculated that it might harbor alien beings who are using it as a base for UFO operations to the nearby planet Earth, no one ever suggested that the Moon itself is a UFO—a spaceship.

It is undoubtedly still difficult for you to believe. I know it is for me. At a glance there does not appear to be anything unusual about this barren, hostile, waterless, airless chunk of rock. However, as we have seen, the Moon is really an invisible island world, for it gives every evidence underneath of being a spaceship.

Frankly, I must confess at times I am overwhelmed from the sheer mind-boggling impact of it all. I find myself asking time and again: How can this be possibly true?

I began with rank skepticism, with a mind closed to even the possibility. I began researching this theory to use the Spaceship Moon idea of the Soviets as a vehicle for a science-fiction novel. Suddenly I began to see that the facts and findings of our lunar program actually backed and even proved the startling, shocking artificial-Moon theory.

Then and only then, as the Apollo findings gathered momentum and compelling evidence poured in, did I finally grudgingly, reluctantly admit it could have validity.

Now the Moon has changed for me—as I imagine it has for you. Now when I see the Moon in our skies I find myself attracted by this strange magnetic mystery world. I stare at her, contemplating how this synchronized satellite has kept this same face gazing down on man's world for eons— as if she were keeping watch on our every move, as if our Spaceship Moon were continually taking the measure of mankind. Wondering, indeed, if she is still inhabited, as our Apollo flights, buzzed as they were by UFOs and plagued by strange radio signals, indicate.

Today when I see the Moon I find myself drawn to her. The strange luminosity of Luna is spellbinding. And the more I dwell on this thundering discovery, the more my own mind becomes numb by the wonderment of it all. I find myself staring at Selene's serene surface and wondering what really is inside this enigmatic ball of rock and metal.

Suddenly this familiar orb which I thought I knew so well seems so different. Its bleak, forbidding, alien-looking surface now appears arresting and awesome—as never before. The Moon for me has become an unearthly experience.

In the privacy of my own world of thoughts I find myself continually dwelling on these staggering ideas. We still find it hard to believe, even after all our research, after all the impressive evidence, that our Moon has not always been a constant companion of Earth but was once a rover in space—an alien world whose masters and creators drove it here across the starlit shores of vast space to lock this spacecraft world in orbit around us.

Many readers have by now bowed before the overwhelming weight of scientific fact. At least, if you are not completely convinced by the compelling evidence marshaled here, you certainly must agree you will never again raise your eyes to the Moon in our heavens without wondering whether or not our nearest neighbor in space is truly a huge alien spacecraft!

I myself, who started out as a complete skeptic, have gradually, through discovery after discovery, fact after fact which turned out to be pieces of the puzzle that fit Spaceship Moon, slowly changed my mind until now I am thoroughly convinced that the Soviet scientists are absolutely correct.

Do I ever doubt the theory? Yes, doubts do crop up from time to time—not factual doubts but mental and emotional misgivings. For it is still yet all so unbelievable. Certainly I have come to reluctantly realize, after all my researching which points unerringly to the truth of the spaceship theory, that I am compelled to accept it. I naturally also felt the urge to tell my fellow humans about this great discovery. At first I hesitated—and more than once. I found I needed the courage of my convictions. I wavered, even as Bishop John Wilkins tells us he did when he too discovered constant references to the "gods" coming from inside the world of the Moon to help mankind.

This Anglican bishop, who authored *Discovery of a New World*—the alien world inside our Moon—admits candidly: "I must needs confess, that I had often thought with myself that it was possible there might be a world in the Moon, yet it seemed such an uncouth opinion that I have durst discovered it, for fear of being counted singular and ridiculous, but afterward having read Plutarch, Galileo and Kepler, with some others, I then concluded that it was not only possible there might be, but probable that there was another habitable world in that Planet."

Like Bishop Wilkins, I hesitated and wavered—for nearly five years—before finally the facts overwhelmed me into actual authorship of my first book, *Our Mysterious Spaceship Moon*. The very idea that the Moon is an alien-created spaceship, the product of alien intelligence, I realize is just on the face of it seemingly much too unbelievable to consider. I realize as I undertake the writing of these works that I too might be in the same boat as other bold original thinkers in this world, and be severely criticized, even labeled a crackpot.

I know that this mind-staggering theory will come under similar fire and I its author might suffer by becoming the target of closed-minded derision. I therefore offer these findings boldly but not without trepidation.

But truth is truth, whatever it might be. Furthermore, my fear does not compare to that of Copernicus and Galileo. I do not have to fear being burned at the stake or even being imprisoned for my findings.

Sincerely, I hope this book will be received not just with open-minded acceptance, but with open-minded consideration and appraisal. Remember, we are all searching for the truth, whatever it might be.

We sincerely hope that scientists who are stumbling around trying to find a solution to the many mysteries of the Moon—to its origins and make-up—seriously look at the evidence and consider the theory.

Although we can hope that even skeptical scientists will open-mindedly give it serious consideration, we are aware that with such a bizarre theory the chances are against it. Most scientists and in fact most people tend to live in their own little world of accepted thought. It is not surprising that human beings who over the years come to accept things as they are interpreted by orthodox scientists have come naturally to look upon the world revolving around Earth (actually, we both revolve around each other) as just an ordinary satellite. Hence, they are not going to be open to accepting—or for that matter even considering—such an unorthodox view.

This is the very reason why revolutionary theories which take a completely different tack from the accepted norm have always had trouble. Not that scientists shouldn't be skeptical. They must be. But still they should not close their minds even to radical possibilities. If they had taken an open-minded attitude hundreds of years ago, neither Copernicus and Galileo would have had such difficulties and harassment.

But, realistically, it is not surprising that men like Copernicus and Galileo ran into a hornet's nest of opposition. After all, human beings had come over the many centuries to look upon their own world as being the center of the universe. When they challenged this concept, claiming that the Sun was not the center of our solar system, it was just too upsetting, too radical for the powers that ran things at that time to accept. In fact, even to tolerate. This naturally provoked fierce opposition and suppression of such a dangerous idea.

Admittedly, it seemed on the surface to be too ridiculous to the average person. Like the Spaceship Moon theory of today, the unbelievable thesis that the Sun, not the Earth, was the center of things appeared preposterous. Why, anyone save a blind man or idiot could see the Sun rose in the east, went across the sky, and set in the west. Anyone could see the Sun went around us!

ANOTHER DANGEROUS IDEA?

Certainly the human race has come a long way since the original suppression of this medieval idea. But accepted, entrenched ideas are difficult to change. Radical, unorthodox concepts such as that of Spaceship Moon will turn out to be too upsetting for the orthodox scientific order of things. Especially if this idea begins to seriously take hold on the mind of the masses, then the scientific world threatened to bow again before established truth will fight a bitter fight once more, we are afraid.

We realize also that making such radical ideas as Spaceship Moon known has proven in the past to be risky business. In ancient Greece five centuries before Christ a philosopher-teacher by the name of Anaxagoras of Clazomenae, a mentor of Socrates, was condemned to death for claiming that the Moon was made of essentially the same material as our Earth. We also know that Galileo was imprisoned and would have been put to the torch if he had not withdrawn his shocking theory about the position of our world in the cosmic scheme of things.

Copernicus, the author of the heliocentric theory, was so afraid that his radical idea might create a storm of controversy resulting in punishment that he held back the publication of his findings until he was on his deathbed.

Of course, scientists today need not worry about such penalties. Neither do I.

Realistically, however, the scientific community as well as the media will ignore this theory—or at best classify this book as science fiction.

A top NASA expert, Edgar Cartwright, in the NASA publication *Apollo Expeditions to the Moon* has made this intriguing observation: "The early telescopes that first revealed the crater-pocked face of the Moon touched off several centuries of speculation about the lunar surface by scientists and science fiction writers alike—it often being unclear who was writing the fiction."

I am sure that although it now appears that science-fiction writers were closer to the truth about the Moon than our scientists, the establishment will collectively hardly take notice of this theory despite the weight of its evidence. While those in the inner sanctum of our space agency will doubtless be disturbed by these revelations, the rest of the scientific community will almost certainly pass it all off as science fiction.

We ardently hope things will be different this time around, but unfortunately they will probably not be.

I can hear the scientific skeptics shouting aloud: *Who are you to stand alone against the scientific world?* Actually, we don't. This, remember, is not our theory—but the brainchild of two Soviet scientific researchers. Also, it now appears that other leading scientists outside the Iron Curtain, including a NASA researcher at the Jet Propulsion Center and an Oxford physicist, agree with it!

However, even if I did stand alone against the world I would be in good company. Remember how often through the history of science such has been the case—Copernicus, Galileo, Pasteur, Kepler, the Wright Brothers, Marconi—all stood alone.

Even our NASA leaders recognized this fact, for in their recent publication *The New Mars* they quote Galileo, one of the "crackpots" whom the scientific and religious establishment opposed and forced to shut up. And it is an interesting quote they pass along: "In questions on science the authority of a thousand is not worth the reasoning of a single individual."

So be it with this theory.

It was Galileo who through his pioneer telescopic studies, as *Science News* put it, "transformed the Moon from a ball of light to a ball of rock." (November 29, 1969.)

Now it looks as though scientists are going to eventually be forced to change their conception of the Moon again—this time from a ball of rock to a world of rocks and metal; from a natural satellite of Earth into a huge hollowed-out spaceship!

How long will it take before this eventually happens? We do not know. But as the great Max Planck, whose quantum theory shook up the scientific world just a few decades ago, once observed: "An important scientific innovation rarely makes its way by gradually winning over and converting its opponents: it rarely happens that Saul becomes Paul. What does happen is that its opponents gradually die out and that the growing generation is familiarized with the idea from the beginning! Another instance of the fact that the future is with youth."

If this day ever dawns, then the world will be in for a revolution—a revolution of changing ideas. For the realization that the Moon is more than just a moon but a world harboring intelligent beings is awesome and disturbing in its implications.

This is a mind-stunning idea, for in facing the reality of

the Moon we may be ultimately facing the reality of ourselves. Our whole new outlook * on our satellite will, we are convinced, lead us to take a whole new look at ourselves—at our origins, our past, our purpose, and, yes, shockingly possible, even our future.

At the beginning we asked you not to accept anything about the Moon but to examine the evidence with an open mind and reach your own conclusions. We have followed the facts, lead where they might. This book has thus been short on speculation, long on evidence and data. And that is the way it should be.

Now an overwhelming array of speculative questions begins to besiege our minds. Who are these intelligent Moon beings? Where do they come from? What do they want with us? (Or should we say, what have they had to do with us, for evidence Morris Jessup cites leads one to believe as he does that they have been around for thousands of years.)

Are their minds like ours (or should we maybe say, are our minds like theirs)? Are they godlike or just beings like ourselves—except, of course, on a much higher level of development? Could they be pure intelligence?

There is no end, seemingly, to the questions that now swirl in one's mind. Questions that hinge and swing around the ultimate questions of man's existence: Where did we come from? Why are we here? Where are we going?

The author does not intend to impinge on the realm of religion. Nor invade the field of the occult. We realize that many people abhor the thought that there may be another form of intelligent life out there—that other intelligences indeed are not only watching and observing us, but are certainly superior to us.

But I am convinced that our minds, shrink as they may from the frightening possible answers to these all-important questions, are so structured and designed that they instinctively hunger and yearn for the truth. Seek the truth whatever it might be. Let the chips fall where they may. And as long as all of us are seeking the truth, whatever it might

---

* With our new Space Age cosmic experiences, more and more people are beginning to spell our Moon and Earth with capital letters; as Buckminster Fuller observes, our new cosmic outlook requires that they should be.

be, there can ultimately be no conflict between religion, philosophy, science, and even the occult. For there can be only one truth—only one true answer to the origin of the Moon, or, for that matter, the origin of man.

Strangely enough, the fact that the Moon is a spaceship seems in our view of the evidence to be the key to a myriad of puzzles perplexing our planet and its inhabitants. Not the least of which are those persistent Unidentified Flying Objects which researchers (and, fortunately, more and more of them scientists!) have found solid evidence for existing not just in our present era but over the long past of mankind—stretching back into the mists of time. What are they and where do they come from? "Spaceships," say the majority of those studying the problem. Although this may be a gross oversimplification—for only God knows what kind of immense intelligences we are dealing with—it would seem to be the logical answer.

In philosophy there is a rule of reason (though not accepted by all) called Occam's Razor that states that if there are a number of equally valid solutions to a philosophical problem, the simplest is probably the correct one. The astronomer Morris Jessup claims that it is also "a basic postulate of science that the simplest explanation is the best." (*The Expanding Case for the UFO.*)

Researchers, I believe—whether they be UFO researchers, philosophical, religious, occult, scientific—will all find that the truth about our Moon is the key to the truth about our world and ourselves. This is Jessup's judgment, too.

We are convinced that this mind-boggling, mind-stunning theory will someday provide a synthesis for all the mysteries swirling around man.

A FIRE TO BE LIGHTED ...

On August 12, 1971, commander David Scott, concluding an Apollo 15 press conference, summed it all up:

"We went to the Moon as trained observers in order to gather data not only with our instruments but with our minds. I'd like to quote a statement from Plutarch which I think expresses our feelings since we've come back: 'The mind is not a vessel to be filled, but a fire to be lighted.'"

We ardently hope that the revelations of this book and the startling solution to the Moon's mysteries will light a fire in many of the brilliant minds that grace our globe. Hopefully, this book will fire these minds not only with the puzzles about our Moon but about man, and impel them to seek to know more about both. Who knows but that someone reading this book will marshal someday the final conclusive evidence, compelling even our secretive government to acknowledge the truth. Maybe then together we will assault the greatest puzzle facing man—himself. For the Moon may be the key even to the mystery of man himself!

In conclusion, the Moon may not be a devastating orb of truth but a fount of revelation about man and his past—the truth of which may set man and his mind free.

Once space enthusiasts described the Moon as a stepping-stone into our universe. It indeed could be just that. And if the Spaceship theory is correct, it could well be *a leaping stone to man's understanding of himself and his own world.*

BIBLIOGRAPHY

Asimov, Isaac. *Asimov on Astronomy*. Doubleday, 1974.
———. *Intelligent Man's Guide to Science*. Basic Books, 1965.
———. *Is Anyone There?* Doubleday, 1956.
Astronauts. *We Seven*. Simon & Schuster, 1962.
Bergier, Jacques. *Extraterrestrial Visitation from Prehistoric Times to the Present*. Henry Regnery, 1973.
Bova, Ben. *The New Astronomies*. St. Martin's Press, 1972.
Cade, Maxwell. *Other Worlds than Ours*. Tapplinger Publications, 1969.
Charroux, Robert. *The Mysterious Past*. Berkley, 1973.
Clarke, Arthur. *The Promise of Space*. Harper & Row, 1968.
———. *Voices in the Sky*. Harper & Row, 1965.
Cole, Dandridge, and Donald Cox. *Islands in Space*. Chilton Books, 1964.
Condon, Edward. *The Scientific Study of Unidentified Flying Objects*. Bantam, 1969.
Cooper, Henry. *Apollo on the Moon*. Dial, 1969.
———. *Moon Rocks*. Dial, 1970.
Corliss, William. *Mysteries of the Universe*. Crowell, 1967.
———. *Strange Universe*. Custom Copy Center, 1975.
Durant, Will. *The Life of Greece*. Simon & Schuster, 1966.
Friedman, Stanton. *UFOs—Myth and Mystery*.
Guest, John, ed. *The Earth and its Satellite*. McKay, 1971.
Hynek, J. Allen. *The UFO Experience*. Henry Regnery, 1972.

Irwin, James. *To Rule the Night: The Discovery Voyage of Astronaut James Irwin.* A. J. Holman, 1973.

Jessup, Morris. *The Case for the UFO.* Citadel Press, 1955.

———. *The Expanding Case for the UFO.* Citadel Press, 1957.

Keyhoe, Donald. *The Flying Saucer Conspiracy.* Holt, 1965.

Leonard, George. *Somebody Else Is on the Moon.* McKay, 1976.

Lewis, Richard. *The Voyages of Apollo.* Quadrangle/New York Times Book Co., 1976.

Magor, John. *Our UFO Visitors.* Hancock House, 1977.

Mansfield, John, ed. *Man on the Moon.* Stein & Day, 1969.

Marsden, B. G., and A. G. W. Cameron. *Earth-Moon System.* Plenum Press, 1966.

Moore, Patrick. *A Guide to the Moon.* W. W. Norton, 1953.

Pensées, editors of. *Velikovsky Reconsidered.* Doubleday, 1976.

Rabanovich, Eugene, ed. *Man on the Moon.* Basic Books, 1969.

Reader's Digest, editors of. *Strange Stories, Amazing Facts.* Reader's Digest Association, 1976.

Riabchiker, E. I. *Russians in Space.* Doubleday, 1971.

Rosenblum, Arthur. *Unpopular Science.* Running Press, 1974.

Ruppelt, Edward. *Report on Unidentified Flying Objects.* Ace, 1964.

Sagan, Carl, and Josif Shklovskii. *Intelligent Life in the Universe.* Holden Day, 1966.

Sendy, Jean. *The Coming of the Gods.* Berkley, 1973.

Shelton, William R. *Winning the Moon.* Little, Brown, & Company, 1970.

Sullivan, Walter. *We Are Not Alone.* McGraw-Hill, 1964.

———, ed. *America's Race for the Moon: Story of Project Apollo.* Random House, 1962.

Stonely, J., and A. T. Lawton. *Is There Anyone Out There?* Warner, 1974.

———. *CETI (Communication with Extra-Terrestrial Intelligence).* Warner, 1976.

Wilkins, H. Percival. *Our Moon.* Frederick Muller, Ltd., 1954.

Wilford, John Noble. *We Reach for the Moon*. W. W. Norton, 1971.
Wilson, Don. *Our Mysterious Spaceship Moon*. Dell, 1975.

ALSO

Lunar Luminescence by M. Sidran and Associates at Grumman Research.

MUFON Symposium by Midwest (Mutual) UFO Network, 1973.

TECHNICAL JOURNALS AND BOOKS

Proceedings of Apollo 11 Lunar Conference. 3 Volumes (2493 pages). Pergamon Press, 1970.

Proceedings of the Second Lunar Conference. 3 Volumes (2818 pages). M.I.T. Press, 1971.

Proceedings of the Third Lunar Conference. 3 Volumes (3263 pages). M.I.T. Press, 1972.

Proceedings of the Fourth Lunar Conference. 3 Volumes (3290 pages). Pergamon Press, 1973.

Proceedings of the Fifth Lunar Conference. 3 Volumes (3134 pages). Pergamon Press, 1974.

Proceedings of the Sixth Lunar Conference. 3 Volumes (3637 pages). Pergamon Press, 1975.

The Moon, An International Journal of Lunar Studies. 13 Volumes. Dordrecht, Holland: D. Reidel Publishing Co., 1969–75.

Handbook of Chemistry and Physics. 54th Edition. Chemical Rubber Co.

NASA DOCUMENTS

Apollo 11: Preliminary Science Report. U.S. Government Printing Office, 1969.

Apollo 12: Preliminary Science Report. U.S. Government Printing Office, 1970.

Apollo 14: Preliminary Science Report. U.S. Government Printing Office, 1971.

Apollo 15: Preliminary Science Report. U.S. Government Printing Office, 1971.

Apollo 16: Preliminary Science Report. U.S. Government Printing Office, 1972.

Apollo 17: Preliminary Science Report. U.S. Government Printing Office, 1973.
Apollo 14: Science at Fra Mauro. U.S. Government Printing Office, 1971.
Apollo 15: At Hadley Base. U.S. Government Printing Office, 1971.
Apollo 16: On the Moon with Apollo 16. U.S. Government Printing Office, 1972.
Apollo 17: On the Moon with Apollo 17. U.S. Government Printing Office, 1972.
NASA's Apollo Expeditions to the Moon. U.S. Government Printing Office, 1975.

POPULAR PERIODICALS CONSULTED

Astronautics
Astronomy
Aviation Today and Space Technology
Chemistry
Detroit Free Press and News
Modern People Press
National Enquirer
National Geographic
New York Times and *New York Times Magazine*
Pensées, Student Academic Freedom Forum Magazine
Physics Today
Popular Science
Saga and *Saga UFO Report*
Science
Science News
Scientific American
Sky and Telescope
Spaceflight: Journal of the British Interplanetary Society
Space World
UFOlogy

PERIODICAL CITATIONS

CHAPTER 1—STRANGE STRUCTURES DISCOVERED ON THE MOON
Argosy, "Mysterious Monuments on the Moon" by Dr. Ivan Sanderson, August 1970.

Washington *Post*, "Six Mysterious Statuesque Shadows Photographed on the Moon by Orbiter," by Thomas O'Toole, November 22, 1966, p. 1.

CHAPTER 2—MYSTERIOUS LIGHTS ON THE MOON!
NASA's Chronological Catalogue of Reported Lunar Events, NASA, July, 1968.
UFOlogy, "Our Mysterious Moon," D. William Hauck, editor, Spring Issue, 1976.
Flying Saucers on the Moon, Riley Crabb, Director of Borderland Sciences Research Assoc., pp. 1–41.

CHAPTER 3—THE LUNAR UFO CONNECTION
Science, "Reopening the Question: *The UFO Experience*," by Dr. Bruce Murray (book review), August 25, 1974, Vol. 177, pp. 688–9.

CHAPTER 4—IS OUR MOON A KIND OF HOLLOWED-OUT SPACESHIP?
Sputnik, "Is the Moon the Creation of Alien Intelligence," by Michael Vasin and Alexander Shcherbakov, July 1970.

CHAPTER 5—MAN'S OWN SPACESHIP WORLD?
Science Fiction Plus: Preview of the Future, "Interstellar Flight," by Dr. Leslie Shepherd, April 1953 (Vol. 1, No. 2), pp. 56–60.

CHAPTER 8—IS OUR MOON HOLLOW?
The Moon, An International Journal of Lunar Science, "Density Within the Moon and Implications for Lunar Composition," by Dr. Sean C. Solomon, Vol. 9, 1974, (pp. 147–165).
Science News Letter: "Moon Like Hollow Sphere," April 22, 1961, p. 244.
"Apollo 12's Moon: Surprises Already," November 29, 1969, pp. 493–4.
"Bonanza from the Highlands," by Everly Driscoll, July 1972 (Vol. 102).
Astronautics, February 1962, p. 225.
Science, November 12, 1971, p. 688.

CHAPTER 9—DOES THE MOON HAVE A CONSTRUCTED OUTER SHELL?

Astronautics and Aeronautics, "The Contending Moon," by Dr. Harold Urey, January 1969.

The Moon, An International Journal of Lunar Science: Bombardment as a Cause of Lunar Asymmetry," by John A. Wood, Vol. 8, 1973 (pp. 73–103).

"The System of Lunar Craters," revised by C. A. Cross, Vol. 3, p. 408ff.

National Geographic, "That Orbed Maiden . . . ," by Kenneth Weaver, February 1969 (pp. 207–230).

The New York Times, "Lunar Churning Surmised from Samples of Rock," by John Noble Wilford, January 7, 1970, p. 32. Also see Walter Sullivan, November 9, 1969.

The New York Times Magazine, "The Moon is a Rosetta Stone," by Dr. Robert Jastrow, November 1969.

Science, "Properties and Composition of Lunar Materials: Earth Analysis," by Edward Schrieker and O. L. Anderson, June 26, 1970.

Scientific American, "The Lunar Soil," by John A. Wood, August 1971 (Vol. 223, No. 2), pp. 14–23.

Science News Journal: "Apollo Rocks Analyzed Tracing the Moon's Origin," August 16, 1969 (Vol. 96), p. 129.

"Migrating Metals on the Moon" (cover title); also "Dating of Moon Samples: Pitfalls and Paradoxes," January 1, 1972 (Vol. 101), pp. 12–13.

"Apollo 15 Data: The Moon's Interior," September 11, 1971 (pp. 107–108).

"Man in the Moon Has Two Faces," December 24, 1966, p. 531.

"Bonanza from the Highlands," by E. Driscoll, July 1, 1972 (Vol. 102), pp. 12–13.

Also see news item *Sc. News Journal,* January 17, 1970, p. 69.

U.S. News and World Report, "As the Moon Yields Its Secrets," January 19, 1970, pp. 28–29.

CHAPTER 10—DOES THE MOON HAVE AN INNER SHELL OR "HULL" OF METAL?

Sputnik, "Is the Moon the Creation of Alien Intelligence?" Michael Vasin and Alexander Shcherbakov, July 1970.

The Moon, An International Journal of Lunar Studies,
> "Moonquakes and Lunar Tectonism," by Gary Latham, M. Ewing, et al., Vol. 7, 1972 (pp. 373–382).
>
> "The Thermal History of the Moon," by C. A. Cross, 1972.
>
> "The Apollo 15 Lunar Heat Flow Measurement," by M. Lanseth, et al., Vol. 4, 1972 (pp. 390–417).
>
> *Fourth Lunar Conference,* "Iron Abundance in the Moon From Magnetometer Measurements," by Dr. Curtis Parkin, Palmer Dyal, and William Daily, Vol. 3 (pp. 2974–2961).

Popular Science, "Our Ideas About the Moon," Dr. Wernher von Braun, January 1972 (pp. 67–68).

Science News: "An Out Of Whack Moon," January 29, 1972 (Vol. 101), p. 73.
> "Possible Lunar Hot Spots," June 12, 1971 (Vol. 96), p. 403.
>
> "Man in the Moon Has Two Faces," December 24, 1966 (Vol. 90), p. 531.

CHAPTER 11—ARTIFICIAL CONSTRUCTION INSIDE THE MOON?

The New York Times, "Seismic Net Set to Find Source of Tremors as Moon Nears," Walter Sullivan, August 4, 1972, pp. 1–8. Also see August 27, 1971.

Science News: "More Light on the Moon," April 3, 1971 (Vol. 99).
> "A Look Inside the Moon," April 7, 1973 (Vol. 103), pp. 228–9.

CHAPTER 12—THE MOON IS NO LONGER A MYSTERY

Chemistry, February 1974.

The Moon, An International Journal of Lunar Studies,

"The Thermal History of the Moon," C.A. Cross, Vol. 4, 1972, pp. 157–158.

The New York Times Magazine, "The Moon Is More of a Mystery Than Ever," by Earl Ubell, April 16, 1972 (pp. 32–3, 50–1).

"The Moon is a Rosetta Stone," by Robert Jastrow, November 9, 1969. Also see *The New York Times* November 16, 1969, and January 7, 1970.

Physics Today, March 1974 (Vol. 27), pp. 44–49.

Sky and Telescope, June 1973.

National Geographic, "That Orbed Maiden . . . ," by Kenneth Weaver, February 1969 (pp. 207–230).

Science, "Is the Moon Hot or Cold?," Dr. D. L. Anderson and T. C. Hanks, Vol. 178, 1972, pp. 1245–9.

Scientific American, "The Magnetism of the Moon," Palmer Dyal and Dr. Curtis Parkin, August 1971 (Vol. 225), pp. 63–73.

Science News: "Moon May Give Evidence on How Stars Are Born," May 15, 1963 (Vol. 87), p. 309.

"Scientists Hold a Landmark Session," January 10, 1970 (Vol. 97), pp. 33–34.

"The Moon's Radioactive Material," April 7, 1973 (Vol. 103), p. 224.

"Another Vote for Moon Water," January 29, 1972 (Vol. 101), p. 93. Also see October 23, 1971.

"How Did the Moon Get Magnetized" (cover title); "That Magnetic Moon: How Did It Get That Way?" by E. Driscoll, May 27, 1972 (Vol. 101), pp. 346–347.

"At the Moon Conference: Consensus and Conflict," June 23, 1971 (Vol. 99), pp. 61–2.

"Possible Observation of Water Vapor on the Moon," October 23, 1971 (Vol. 100) p. 277.

Third Lunar Conference: "Water Vapor, Whence Comest Thou?" by Dr. J. W. Freeman, A. Hill, R. Vandrak, Vol. 3, p. 227.

CHAPTER 13—THE ALL-COMPREHENSIVE SPACESHIP MOON THEORY!

The Moon: An International Journal of Lunar Studies, "Origin and Evolution of the Earth-Moon System," by H. Alfren and G. Arrehenius, Vol. 5, 1972, pp. 211–229.

Science News: "A Look Inside the Moon," by E. Driscoll, April 7, 1973 (Vol. 103), pp. 228–9.

"Apollo Returns," August 2, 1969 (Vol. 96), pp. 95–96.

CHAPTER 14—SOMEBODY IS INSIDE THE MOON!

Scientific American, "Carbon Chemistry of the Moon," by G. Englinton, James R. Maxwell, and Colin T. Pillinger, October 1972 (Vol. 227), pp. 80–90.

Science, "Reopening the Question: *The UFO Experience*," by Dr. Bruce Murray (book review), Vol. 177, p. 688.

Missiles and Rockets, August 10, 1964, p. 72.

Detroit Free Press, "Mystery Force Pesters Apollo Station" (*New York Times* News Service), April 22, 1976 (p. 8-B).

CHAPTER 15—THE MULTIPLYING MYSTERIES OF THE MOON

Science Digest, "Is There a Tunnel on the Moon?" November 1952 (Vol. 32), p. 70.

CHAPTER 18—THE MYSTERIOUS ALIEN SATELLITE OF OUR MOON!

Spaceflight, The Journal of the British Astronomical Society, "Space Probe From Epsilon Bootis," by D. A. Lunan, April 1973, pp. 122–133.

Time, "Message From a Star . . . ," April 9, 1973, pp. 59–60.

And Selected From Sphere's Non-Fiction List

MASTERS OF THE WORLD

'Enigmatic objects fly across our sky; monuments whose purpose is unknown to us stand on the surface of our land, and beneath it are buried structures belonging to no known civilization. Mystery is all around us and neither our science nor our history can give an answer to it. In spite of everything, and in the face of opposition, silence or disapproval from those who do not want the veil to be lifted, Man tries to break open the door that leads to knowledge. Documents are speaking, initiates are breaking the seals of tablets hidden in sanctuaries — Man will soon know much more about his unknown past . . .'
Robert Charroux in MASTERS OF THE WORLD

Among Charroux's shaking disclosures are: —
* Proof that a Universal Deluge — with waves of six thousand feet — did indeed occur in ancient times
* The true facts behind the Miracle at Fatima
* The fascinating history of the Rosicrucians, for many centuries the most carefully guarded secret society in the world
* The secret powers of jade — and the enigma of The Man in the Jade Mask . . .
* The strange stone discs of Tibet with their awesome message of spacecraft visiting Earth in remote prehistory

These and many more such revelations will make you change the way you think about history — and your own lives . . .

COSMOLOGY   0 7221 2271 3   £1.25

ALTERNATIVE 3

by Watkins, Ambrose & Miles

THE MOST ASTOUNDING AND FRIGHTENING CONSPIRACY EVER

Research for what was originally intended as a straightforward TV documentary on the scientific 'Brain Drain' from Britain revealed some extremely disturbing things:

* Many people joining the Brain Drain are vanishing off the face of the Earth — literally
* Earth will soon be unable to support life: our climate's recent strange behaviour is only a warm-up for the cataclysms to come
* The super-powers have been working secretly together in space for *decades*
* Government agencies are kidnapping ordinary people and turning them into mindless slaves by advanced brainwashing methods
* Astronauts' reports of strange things they saw on the Moon have been suppressed
* Ultra-secret joint US/USSR conferences are held each month in a submarine beneath the Arctic ice-cap

And this was just the tip of the iceberg. Behind these and many more sinister features lurks the top secret operation known as ALTERNATIVE 3 — an international government conspiracy so monstrous that the human mind can scarcely grasp its true enormity. This courageous book goes beyond even the ground-breaking TV exposé to reveal the full awesome horror of ALTERNATIVE 3 . . .

OCCULT/COSMOLOGY   0 7221 1145 2   95p

SOMEONE ELSE IS ON OUR MOON
by George H. Leonard

AN EXCAVATING MACHINE AS BIG AS A CITY — ON THE SURFACE OF THE MOON! AND THAT'S JUST ONE OF GEORGE H. LEONARD'S SENSATIONAL, SHOCKING REVELATIONS.

Few people noticed the secret code words used by astronauts to describe the Moon.
Until now, few knew about the strange moving lights they reported.
Or were aware of the huge mechanical contrivances seen working in the craters of the Moon.

George H. Leonard fought through the official veil of secrecy and studied thousands of NASA photographs, talked candidly with dozens of officials from NASA and listened to hours of astronauts' tapes. Here he presents the stunning and inescapable conclusion of his work:

THE SPACE AGENCY — AND MANY OF THE WORLD'S TOP SCIENTISTS — HAVE KNOWN FOR YEARS THAT THERE IS INTELLIGENT LIFE ON THE MOON.

COSMOLOGY   0 7221 5486 0   £1.25

And from Abacus Books

MARS AT LAST!
by Mark Washburn

MARS GIVES UP HER SECRETS

When the *Viking* Lander touched down on the surface of Mars in the autumn of 1976, Man had finally reached the planet that had fascinated him for centuries. In MARS AT LAST! Mark Washburn brings together all that Man has dreamt and learned about the Red Planet, from the war-symbolism of the ancient Greeks, Persians and Romans, through the flights of fancy of such writers as H. G. Wells and Robert Heinlein, right up to full details of the discoveries made by the *Viking* Missions.

SCIENCE   0 349 13591 6   £1.75